# Genomic Medicine in Emerging Economies

**Translational and Applied Genomics Series**

# Genomic Medicine in Emerging Economies

## Genomics for Every Nation

Edited by

**Catalina Lopez-Correa**

Genome British Columbia, Vancouver, BC, Canada

**George P. Patrinos**

University of Patras School of Health Sciences, Patras, Greece

United Arab Emirates University, Al-Ain, United Arab Emirates

ACADEMIC PRESS

An imprint of Elsevier

Academic Press is an imprint of Elsevier
125 London Wall, London EC2Y 5AS, United Kingdom
525 B Street, Suite 1650, San Diego, CA 92101, United States
50 Hampshire Street, 5th Floor, Cambridge, MA 02139, United States
The Boulevard, Langford Lane, Kidlington, Oxford OX5 1GB, United Kingdom

**Notices**
Knowledge and best practice in this field are constantly changing. As new research and experience broaden our understanding, changes in research methods, professional practices, or medical treatment may become necessary.

Practitioners and researchers must always rely on their own experience and knowledge in evaluating and using any information, methods, compounds, or experiments described herein. In using such information or methods they should be mindful of their own safety and the safety of others, including parties for whom they have a professional responsibility.

To the fullest extent of the law, neither the Publisher nor the authors, contributors, or editors, assume any liability for any injury and/or damage to persons or property as a matter of products liability, negligence or otherwise, or from any use or operation of any methods, products, instructions, or ideas contained in the material herein.

**Library of Congress Cataloging-in-Publication Data**
A catalog record for this book is available from the Library of Congress

**British Library Cataloguing-in-Publication Data**
A catalogue record for this book is available from the British Library

ISBN: 978-0-12-811531-2

For information on all Academic Press publications visit our website at
https://www.elsevier.com/books-and-journals

  Working together
to grow libraries in
Book Aid
International developing countries

www.elsevier.com • www.bookaid.org

*Publisher:* John Fedor
*Acquisition Editor:* Peter B. Linsley
*Editorial Project Manager:* Leticia Lima
*Production Project Manager:* Punithavathy Govindaradjane
*Designer:* Matthew Limbert

Typeset by Thomson Digital

# Contents

# List of Contributors

**Viviana Ada Bernath**   Laboratorio Genda sa, Buenos Aires, Argentina

**Fahd Al-Mulla**   Kuwait University, Health Sciences Center, Kuwait City, Kuwait

**Carolina Alvarez**   Department of Cell and Molecular Biology, Faculty of Biological Sciences, Pontificia Universidad Católica de Chile, Santiago, Chile

**Karin Alvarez**   Laboratory of Oncology and Molecular Genetics, Coloproctology Unit, Clinica las Condes, Santiago, Chile

**Angela Brand**   University of Maastricht, Institute of Public Health Genomics, Maastricht, The Netherlands

**Alicia Maria Cock-Rada**   Instituto de Cancerología SA, Medellin, Colombia; Universidad de Antioquia, Medellin, Colombia

**Pilar Carvallo**   Department of Cell and Molecular Biology, Faculty of Biological Sciences, Pontificia Universidad Católica de Chile, Santiago, Chile

**David N. Cooper**   Institute of Medical Genetics, Cardiff University, Cardiff, United Kingdom

**Debora M. de Miranda**   Universidade Federal de Minas Gerais, Belo Horizonte, Brazil

**Vajira H.W. Dissanayake**   Human Genetics Unit, Faculty of Medicine, University of Colombo, Colombo, Sri Lanka

**Vita Dolzan**   University of Ljubljana, Ljubljana, Slovenia

**Mev Domínguez**   Department of Tumor Biology, Institute for Cancer Research, Oslo University Hospital, Oslo, Norway

**Camilo A. Espinosa Jovel**   Hospital de Kennedy, Bogotá, Colombia

**Diego A. Forero**   Universidad Antonio Nariño, Bogotá, Colombia

**Paolo Fortina**   Thomas Jefferson University, Kimmel Cancer Center, Philadelphia, PA, United States

**Yeimy González-Giraldo**   Pontificia Universidad Javeriana, Bogotá, Colombia

**Federico Innocenti**   Institute of Pharmacogenomics and Individualized Therapy, University of North Carolina, Chapel Hill, NC, United States

**Theodora Katsila**   University of Patras School of Health Sciences, Patras, Greece

**Ming T.M. Lee**   Genomic Medicine Institute, Danville, PA, United States; RIKEN Center for Integrative Medical Sciences, Yokohama, Japan

**Catalina Lopez-Correa**   Genome British Columbia, Vancouver, BC, Canada

**Sandra Lopez-Leon**   Novartis Pharmaceuticals Corporation, East Hanover, NJ, United States

**Milan Macek Jr.**   Charles University Prague, Institute of Biology and Medical Genetics, Prague, Czech Republic

**Konstantinos Mitropoulos**   The Golden Helix Foundation, London, United Kingdom

**Christina Mitropoulou**   The Golden Helix Foundation, London, United Kingdom

**Anastasios Mpitsakos**   University of Patras School of Health Sciences, Patras, Greece

**Carlos Andres Ossa Gomez**   Clinica Las Américas, Medellin, Colombia

**George P. Patrinos**   University of Patras School of Health Sciences, Patras, Greece; United Arab Emirates University, Al-Ain, United Arab Emirates

**Mariana Herrera Piñero**   Banco Nacional de Datos Genéticos, Buenos Aires, Argentina

**Barbara Prainsack**   King's College London, London, United Kingdom

**Nirmala D. Sirisena**   Human Genetics Unit, Faculty of Medicine, University of Colombo, Colombo, Sri Lanka

**Alessio Squassina**   University of Cagliari, Cagliari, Italy

**Domenica Taruscio**   National Centre for Rare Diseases, Istituto Superiore di Sanità, Rome, Italy

**Hermes Urriago**   Universidad Antonio Nariño, Bogotá, Colombia

**Ron H. van Schaik**   Erasmus University Medical Center, Rotterdam, The Netherlands

**Effy Vayena**   University of Zurich, Epidemiology Biostatistics and Prevention Institute, Zurich, Switzerland

**Athanassios Vozikis**   University of Piraeus, Piraeus, Greece

**Wei Wang**   Edith Cowan University, Perth, WA, Australia; Capital Medical University, Beijing; Taishan Medical University, Taian, China

**Marc S. Williams**   Genomic Medicine Institute, Danville, PA, United States

**Bauke Ylstra**   VU University Medical Center, Amsterdam, The Netherlands

# Genomic Medicine in Developing and Emerging Economies: State-of-the-Art and Future Trends

**Catalina Lopez-Correa\*, George P. Patrinos\*\*,†**
*\*Genome British Columbia, Vancouver, BC, Canada; \*\*University of Patras School of Health Sciences, Patras, Greece; †United Arab Emirates University, Al-Ain, United Arab Emirates*

## INTRODUCTION

Genomic medicine refers to those medical interventions that employ the information encoded in an individual's or patient's genetic material to amend medical interventions, such as drug prescription and preventive medical testing, and guide lifestyle decision-making, aiming to improve the overall quality of life. According to Hippocrates of Kos (460–370 BCE) "…it is more important to know what kind of person suffers from a disease than to know the disease a person suffers," highlighting the fact that the unique genetic composition of every individual directly impacts on his health. Hence two individuals suffering from the same disease and treated with the same drug and drug dose might present with different treatment efficacy, from (1) being fully treated, (2) to not being treated at all, or worse, (3) to developing mild to serious and often life-threatening adverse drug reactions. Ineffective treatment can lead to adverse reactions and toxicity events in certain patients, which not only causes the deterioration of their overall health status but also leads to increased treatment costs. Greater knowledge regarding the genetic variation between individuals may potentially help in selecting the appropriate treatment. Because of the limited resources in health care systems, it is necessary to consider the justification of current spending and future investment in public health care. Genomic medicine is a subset of precision (or personalized) medicine, which constitutes an emerging approach for disease treatment and prevention that takes into account individual variability in genes, environment, and lifestyle for each person. This approach allows clinicians to predict more accurately what treatment

Genomic Medicine in Emerging Economies. http://dx.doi.org/10.1016/B978-0-12-811531-2.00001-1

and prevention strategies should be followed in every group of people. To date genomic medicine interventions are being implemented in routine clinical practice in various parts of the world to a greater or lesser extent.

However, implementation of genomic medicine has not always advanced at a uniform pace in all parts of the world. In some developed countries, such as Canada, the United States, and those in northern Europe, genomic medicine is being implemented at a more rapid pace, which results from large investments in genomics implementation projects, state-of-the-art infrastructure, sufficient genetic literacy of healthcare professionals, established reimbursement policies for genetic testing, and legislation governing the use of genetic tests. Whereas, in the majority of resource-limited environments in South East Asia, the Middle East and Latin America, the pace of genomic medicine implementation is much slower; although, there are several examples of successful implementation of genomic medicine that are confined to certain conditions that are mostly prevalent in a particular region, but may constitute a significant health and/or economic burden for these countries. To a certain extent, genomic medicine initiatives in developed countries have been driven by countrywide strategies, whereas the use of genomics and implementation of genomic medicine practices in emerging economies has been much more targeted to specific local needs or opportunities.

## WHAT ARE EMERGING ECONOMIES?

Countries are generally classified into developed and developing countries by using different classification systems. There are many factors used to rank countries, such as national income, human development, and growth rate. For instance, the United Nations (UN) uses the UN Human Development Index, which is a summary measure of the average achievements in key aspects of human development; the World Bank (WB) uses gross national income per capita for this form of classification.

These classifications are used to assess how a country is progressing from an economic and social perspective, as well as what needs remain unmet, such as funding or long-term strategies related to health, education, and scientific research. Indeed, the fields of health, education, and scientific research need attention in the developing countries since they are tightly connected with human development and the country's national income. These are the areas that are typically underrepresented and/or poorly funded and that represent the most evident difference between developed countries and emerging economies.

According to a publication in "The Balance" in 2017 (Christy, 2017), emerging markets, also known as emerging economies or developing countries, are nations that are investing in more productive capacity. They are moving away from their traditional economies, which have relied on agriculture and the export of raw materials. Leaders of developing countries want to create a better quality of life for their people. Therefore, they are rapidly industrializing and adopting a free market or mixed economy. Emerging markets are important because they drive growth in the global economy.

In this book, we use the notions "developing country," "emerging economies," and "resource-limited settings" to denote an environment in which (1) resources assigned for genomics research and infrastructure are scarce, (2) access to genomic knowledge and information is restricted to only small groups of highly qualified individuals, (3) clinical implementation of genomics is limited to a handful of examples, (4) genomics education of healthcare providers is poor or nonexistent, and/or (5) collaborative opportunities with renowned institutions are rare in most cases because of geographical, societal, political, or economic barriers.

## ARE DEVELOPING COUNTRIES AND EMERGING ECONOMIES INVESTING IN GENOMICS?

One of the challenges or criticisms facing the development and application of new genomic technologies is the perception that these technologies will only benefit the more developed and rich countries and that emerging economies will not be able to afford the cost or will not have the necessary infrastructure to successfully benefit from these new technologies. Emerging economies face specific challenges with respect to the support genomics research. Compared with developed countries, most developing countries lack the qualified personnel, infrastructure, and research centers that could spearhead the generation of new knowledge by training and educating young scientists, medical students, clinicians, etc.

Developed countries in general devote an important portion of their GDP to research and development expenditures. According to the WB this was the case in 2015 for Germany (2.88% GDP), Denmark (3.01% GDP), Finland (2.90% GDP), and Sweden (3.26% GDP). On the other hand, the research and development investment in 2015 for developing countries was much lower, as illustrated by countries like Colombia (0.24% GDP), Chile (0.38% GDP), India (0.63% GDP), and Thailand (0.63% GDP) (World Bank, 2015).

An interesting observation is that these limited investments in research could either lead to a very limited and poor implementation of clinical genomic practices or could, in some cases, lead to a successful more targeted development and application of these new technologies.

## ARE THE APPLICATIONS OF GENOMICS TECHNOLOGIES LIMITED TO HUMAN HEALTH?

Another important aspect that is relevant for developing countries and emerging economies is that aside from some of the well-known applications of genomics in human health, genomics and other "omics" disciplines have proven to be valuable tools that can be effectively applied to many sectors of the economy, including agriculture and the food industry. In fact, genetics and genomics has been successfully used in developing countries in the agrifood sector, in particular, in the crop and livestock industries (Bohra et al., 2014; Piccoli et al., 2017). The ability to become proficient and deploy omics offers opportunities for emerging economies to advance knowledge in many sectors of the economy as well as improve health risk identification, diagnoses, treatment, and prevention of human diseases. The advantage of genomic technologies is that the same infrastructure (e.g., sequencers, etc.) that can be used to develop projects in human health can also be used in other sectors of the economy. Some developed countries, like Canada, have advanced a multisector model, where genomic technologies are used to promote social and economic development (Jimenez-Sanchez, 2015). Through a coordinated investment effort in Canada, genomics has become a major driving force that is helping generate new products and services that are addressing global challenges affecting the world's population. These innovations are helping address health care, food safety, and climate change issues in a more sustainable manner. Many of these bioeconomy genomic innovations could eventually be applied in a cost-effective manner in developing countries or emerging economies (see also Chapter 10).

## WHY SHOULD WE USE GENOMICS IN DEVELOPING COUNTRIES?

When one takes into account that around 85% of the world's population lives in developing/resource-limited countries, it becomes apparent that the issue of implementing genomic medicine practices in these settings is crucial. These countries are all in desperate need of more effective and precise health care systems that will help all patients get a better diagnosis, more targeted treatment, and, most importantly, shift from a disease-centered to a prevention-focused system. As described in the following chapters, many countries are already exploring different ways of implementing genomics and applying "omics" technologies in their health care systems. The pace of implementation of genomic medicine varies from country to country, depending on a number of different parameters. These parameters include barriers—arising mainly from the lack of highly qualified personnel—resources and infrastructure, poor technology and knowledge transfer strategies, and limited genomics knowledge among healthcare providers. Moreover, there is occasionally misplaced skepticism regarding the potential benefits that this new discipline has to offer, as well as the many perceived challenges of applying and implementing these new technologies in resource-limited health care systems (Forero et al., 2016).

The clinical implementation and general use of genomics technologies in the developing world faces many different challenges, among these are the lack of genomic laboratory infrastructure and the lack of a coordinated effort to impart the necessary knowledge, skills, and attitudes on the part of the healthcare workforce to successfully implement genomic medicine. Despite these barriers, there have been several notable examples of successful implementation of genomic medicine projects in resource-limited settings across different continents (Mitropoulos et al., 2017, 2015; Tekola-Ayele and Rotimi, 2015). Many of these examples are discussed in detail in the chapters that follow, and some of these are summarized here by way of introduction.

## Africa

Even though not covered in this volume, there are some good and encouraging initiatives that indicate the potential impact medical genomics could have in the African continent. Before, and even up to a decade after, the Human Genome Project, genetic research in Africa has progressed very slowly with very limited initiatives and limited capacity building. However, in recent years this trend has been rapidly changing, with the advent of large-scale and international initiatives such as the Human Heredity and Health in Africa (H3Africa) initiative and other international initiatives (Coles and Mensah, 2017; Karikari et al., 2015). The H3Africa initiative in particular aims to facilitate a contemporary research approach to the study of genomics and environmental determinants of common diseases, with the goal of improving the health of African populations. To accomplish this, the H3Africa initiative aims to contribute to the development of the necessary expertise among African scientists, and to establish networks of African investigators. The initiative is jointly funded by the Wellcome Trust (the United Kingdom) and the National Institutes of Health (the United States) and is addressing the following issues:

- Ensuring access to relevant genomic technologies for African scientists.
- Facilitating integration between genomic and clinical studies.
- Facilitating training at all levels and particularly in training research leaders.
- Establishing necessary research infrastructure.

But to fulfill this vision, there is a clear need to perform more research on African genomes and to develop local capacity (Mlotshwa et al., 2017). Most genomic studies that have generated catalogs of whole genome sequences so far have focused on white people of European descent. It is estimated that only

3% of global genome-wide association studies—which link genetic traits to patterns in health, disease, or drug tolerance—has been performed on Africans, compared with 81% on people of European ancestry (Nordling, 2017).

## Asia

One of the challenges described in this book regarding the implementation of genomic technologies in Asia is the financial coverage of potential clinical genomic tests. Key questions such us who pays for the tests and who establishes the test pricing have not been addressed in most of Asian countries. Countries like Sri Lanka are developing and using genomic tests, but their national health services do not reimburse the cost of genomic testing and there is no coverage for genomic tests by insurance companies (see Chapter 2). As such, patients have had to bear the burden of financing these tests out of their own pockets. A particularly important task, aside from developing research capacity, has been to convince national governments that genomic medicine is important and that allocating funding for genomic medicine infrastructure is vital for providing affordable genomic medicine services (Sirisena et al., 2016). Another important challenge has been the capacity to train a genomics workforce with expertise in integrating genomic data into clinical delivery services and provides genomic education for both clinicians and patients. As in many other jurisdictions, including developed countries, Sri Lanka has encountered many barriers when trying to implement genomics in the clinic. Some of these include setting up regulatory frameworks to oversee the ethical conduct of genomic research, training genomic scientists, and providing access to both data and genetic resources, to provide avenues for integration of genomic data into clinical practice (see Chapter 2). Issues around data management have been particularly challenging, in particular the sharing of de-identified data to foster multidisciplinary collaborations for genomics research and services and the building of public trust and confidence in genomics research, genetic data sharing, and contribution of samples and data to biobanks (Pang, 2013).

Another example discussed in this book, as far as implementation of genomic medicine in Asia is concerned, is the genomics-related public health programs and services in China (Chapter 3), in particular the role in Chinese health care played by prenatal diagnosis, newborn screening, and genetic testing for rare disease (Zheng et al., 2010). China has developed several initiatives in these areas. Some of the challenges described in this book are the absence of genomic education of health care providers and the lack of availability of genetic counselors. Other interesting aspects are the concepts around the integration and interaction of genomics and traditional Chinese medicine (Wang and Chen, 2013). Some examples discussed are the application of genomics theory to support acupuncture practice and the application of genomic medicine to investigate herb-drug interactions. Finally, the opening of the China

National GeneBank in 2016, which is supported by the world's most advanced and sophisticated high-throughput sequencing and bioinformatics capacity, will provide an unprecedented opportunity to conquer diseases through genomic tools.

## Latin America

Like other emerging economies in the world, several countries in Latin America have been developing small and targeted initiatives in medical genomics that are now following the path to clinical implementation. Genetic screening of cancer genes in Latin American countries have been performed for the last 10–15 years and have been mainly focused on breast and colorectal cancer, permitting advances in clinical decisions based on genetic diagnosis in the countries were these studies have been performed (Sanchez et al., 2011). The size, scope, and extent of these studies varies widely from country to country. These studies are small and only represent a limited number of the population but are already an initial step toward the use of genomics technologies. In addition, ancestry determination in cancer patients is highly relevant in order to determine the influence of ethnic origins in the development of cancer (Alvarez et al., 2017). To increase knowledge and expertise in cancer genetics, a significant increase in funding for research grants, training of young scientists in genetics, installation of PhD programs in genetics and genomics, and an increased overall investment and interest from local government in Latin American countries is necessary (Chapter 4). An illustration of the correlation between investment in science and the implementation of clinical genomics practices can be found in Brazil, where strong investments in science have led to the very advanced "omics" programs (Wünsch-Filho et al., 2006). Similar examples from the transition of genomics knowledge from academia to industry and the establishment of corporate entities with genomics-related activities are reported in Argentina (see Chapter 7).

An important aspect explored in this book is the benefits and challenges of collaborations between developed countries and developing or emerging economies. After many years of collaborations that have been mostly controlled by developed countries, in particular as they relate to the overall benefit and outcomes of research, there is now a new tendency in which developing countries are promoting more balanced collaboration projects. Collaborations between Latin American scientists and scientists from countries with a more developed science are very welcome and necessary (see Chapter 4); however, the collaborations are now aiming at keeping the samples in the country of origin (Zawati et al., 2014).

Aside from extensive work being done in cancer, this book also presents work being done in the area of molecular genetics of psychiatric disorders in Latin America. Some of this work focuses on the replication of previous findings from candidate genes identified in other populations (see Chapter 6). However, most often these studies do not confirm findings previously reported in other populations (Gatt et al., 2015). One of the possible reasons for the inconsistencies found may result from the heterogeneity of the populations or from the differences in the genes involved in the pathophysiology of these disorders. This is the case for other complex and rare diseases where variants found in Caucasian populations are not always replicated in Latin American, Asian, or African countries (Hindorff et al., 2017).

In Latin America, as in many other emerging economies, the strategic implementation of genomic technologies is a fundamental need. It is crucial to continue the work of breaking down the barriers to the use and implementation of clinical genomics to ensure that all populations benefit from these great advances.

## RETHINKING THE INNOVATION MODEL FOR GENOMICS WITHOUT BORDERS

The current innovation model that has shaped scientists' and researchers' approach to scientific breakthroughs is linear, initially starting from discovery work and subsequently proceeding, in a linear fashion, toward translational research, in various phases, concluding with the clinical implementation of the science and, reciprocally, diffusion of innovation and capacity building. However, this model does not take into account that innovations are initiated from discovery science, which in some cases may not be realistic in countries with limited resources. Based on this notion and on the fact that scientists and researchers may not only commence innovations upstream but also "midstream," a new model of innovation, also known as "the Fast-Second Winner" model, was proposed, aiming for long-term development.

In particular, the Fast-Second Winner model takes into account the different public health priorities and disease burdens of individual countries and encourages midstream innovations, based on highly relevant diagnostic biomarkers that differ from country to country. One of the key advantages of initiating innovation midstream is that it can allow innovators to learn from other innovators' mistakes upstream in discovery work and, hence, raise the chances of success for translational and (clinical) implementation science, particularly when resources, in their own case are limited. Contrary to the linear innovation model, which entails horizontal investment in all aspects of innovation from discovery to translation to implementation work in a given country, the proposed à la carte model of global innovation allows different developing countries to embark on multiple entry points into the global genomics innovation ecosystem depending on their own needs, whether or not extensive discovery infrastructure is already in place. Not only is this latter model cost-effective but it also serves the unique needs of each country. In fact, some of the examples presented in this book are small illustrations of the "Fast-Second Winner" model discussed here.

To implement this model widely, there is a need to promote the establishment of innovation observatories, which can best identify a given country's public health needs, such as emerging candidate markers or drug-genomic marker combinations. It would then be possible to make suggestions and provide recommendations to a given country.

## CONCLUSIONS AND FUTURE PERSPECTIVES

Implementation of genomic medicine in developing countries and resource-limited environments can be a rather cumbersome and lengthy process, which can deprive population groups residing in these countries of the benefits that genome-guided clinical interventions can offer. Given the scarcity or even lack of resources in certain developing countries, genomic medicine interventions can only be implemented in these environments under certain conditions.

There are different reasons for this. First, generation of primary research data in research establishments of low-resource environments can be a rather daunting task. As such, and in order to advance genomics research in these countries, strategic partnerships should be sought with other research entities in developed countries, which are likely to create benefits for all parties involved (Cooper et al., 2014). In this case, not only will developing countries benefit from the various training opportunities that will be created on top of knowledge transfer, but developed countries may benefit too, through participation in multicenter projects to study cases and individuals with unique clinical features and/or rare diseases (Fig. 1.1) (Manolio et al., 2015).

Second, it would be highly beneficial for developing countries and low-resource environments to prioritize their research efforts toward implementing those actionable genomic medicine interventions that are relevant to their own needs and that may differ from other parts of the world, including but not limited to pharmacogenomic tests, genetic testing for highly prevalent inherited conditions, and application of genomics technologies to improve the care of cancer patients. To a certain extent, all of these initiatives require the generation of country-, population-, or even region-specific data. As part of developing these local initiatives, it is also important to perform cost benefit analysis and health economic studies that could then help demonstrate the concrete benefits for local health care systems (see Chapter 8).

Third, it is reasonable that policymakers invest a significant amount of effort to expanding the public health aspects of genomic medicine interventions, such

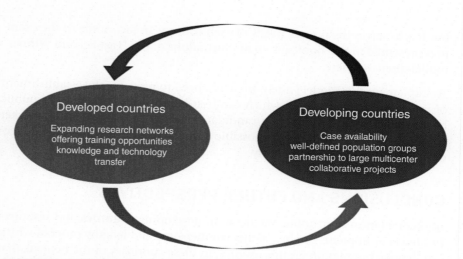

**FIGURE 1.1** Schematic drawing depicting the tangible benefits from the collaboration between developed and developing/resource-limited countries in the field of genomic medicine (see text for details).

as (1) inquiring of the stakeholders' opinions and stance related to genomics; (2) legislating new measures to establish a legal framework for the provision of genetic services, so that not only quality of genetic services is ensured but also so that the general public and the patients are safeguarded; and (3) adopting guidelines for the pricing and reimbursement of genetic tests based on related cost-effectiveness analyses.

Fourth, investment in the ongoing genomics education of healthcare professionals and biomedical scientists, both at the undergraduate level (by the harmonization of genomics education in university curricula) and at the graduate level (with training activities and specialized conferences focused on genomic medicine–related themes) (see Chapter 9), should be set as one of the top priorities, particularly for developing countries and low-resource environments. Providing such education would be much easier than, for example, acquiring expensive genomic infrastructure, and would surly come with a short-term return on such an investment. Given the low cost, and even lower in the next years, of genome sequencing and genomics technologies in general, the real bottleneck now is not the generation of genomics data, but the interpretation and clinical delivery, which makes even more important the development of highly qualified personnel in developing countries.

Last but not least, it is imperative that strong multinational consortia and research networks are established at a regional level or, optimally, internationally to ensure that the efforts to implement genomic medicine in developing countries benefit the respective populations. Networks like the Genomic

Medicine Alliance (www.genomicmedicinealliance.org; see Chapter 10) or the Global Genomic Medicine Collaborative (www.g2mc.org) have already been formed, providing several successful examples from the very early years of their establishment, and demonstrating the tangible benefits that can arise from international cooperation and synergies among developing nations in order to achieve better health outcomes.

We strongly believe that the examples of successful implementation of genomic medicine described in this book, although not exhaustive, could serve as examples to be readily replicated in other countries, such as in the Middle East and Europe (Balkan and Eastern European countries), with a view of expediting their transition to genomic medicine by harmonizing their strategies and policies with those of various other national health care systems that already enjoy the tangible benefits of genomics. Clearly, the inclusion of other omics disciplines (e.g., proteomics, transciptomics, epigenomics, metabolomics, etc.) into precision medicine will further contribute to these developments.

## Acknowledgment

We thank all the authors of the chapters of this book for their contributions, highlighting the importance of implementing genomic medicine in developing nations.

## References

Alvarez, C., Tapia, T., Perez-Moreno, E., Gajardo-Meneses, P., Ruiz, C., Rios, M., Missarelli, C., Silva, M., Cruz, A., Matamala, L., Carvajal-Carmona, L., Camus, M., Carvallo, P., 2017. BRCA1 and BRCA2 founder mutations account for 78% of germline carriers among hereditary breast cancer families in Chile. Oncotarget 8 (43), 74233–74243.

Bohra, A, Pandey, M.K., Jha, U.C., Singh, B., Singh, I.P., Datta, D., Chaturvedi, S.K., Nadarajan, N., Varshney, R.K., 2014. Genomics-assisted breeding in four major pulse crops of developing countries: present status and prospects. Theor. Appl. Genet. 127 (6), 1263–1291.

Christy, J., 2017. Top Emerging Market Economies in the Global Economy. Available from: https://www.thebalance.com/top-emerging-market-economies-1979085.

Coles, E., Mensah, G.A., 2017. Geography of genetics and genomics research funding in Africa. Glob. Heart 12 (2), 173–176.

Cooper, D.N., Brand, A., Dolzan, V., Fortina, P., Innocenti, F., Michael Lee, M.T., Macek, Jr., M., Al-Mulla, F., Prainsack, B., Squassina, A., Vayena, E., Vozikis, A., Williams, M.S., Patrinos, G.P., 2014. Bridging genomics research between developed and developing countries: the genomic medicine alliance. Pers. Med. 11 (7).

Forero, D.A., Wonkam, A., Wang, W., Laissue, P., López-Correa, C., Fernández-López, J.C., Mugasimangalam, R., Perry, G., 2016. Current needs for human and medical genomics research infrastructure in low and middle income countries. J. Med. Genet. 53 (7), 438–440.

Gatt, J.M., Burton, K.L., Williams, L.M., Schofield, P.R., 2015. Specific and common genes implicated across major mental disorders: a review of meta-analysis studies. J. Psychiatr. Res. 60, 1–13.

Hindorff, L.A., Bonham, V.L., Brody, L.C., Ginoza, M.E.C., Hutter, C.M., Manolio, T.A., Green, E.D., 2017. Prioritizing diversity in human genomics research. Nat. Rev. Genet.

Jimenez-Sanchez, G., 2015. Genomics: the power and the promise. Genome 58 (12), vii–x.

Karikari, T.K., Quansah, E., Mohamed, W.M., 2015. Developing expertise in bioinformatics for biomedical research in Africa. Appl. Transl. Genomics 6, 31–34.

Manolio, T.A., Abramowicz, M., Al-Mulla, F., Anderson, W., Balling, R., Berger, A.C., Bleyl, S., Chakravarti, A., Chantratita, W., Chisholm, R.L., Dissanayake, V.H., Dunn, M., Dzau, V.J., Han, B.G., Hubbard, T., Kolbe, A., Korf, B., Kubo, M., Lasko, P., Leego, E., Mahasirimongkol, S., Majumdar, P.P., Matthijs, G., McLeod, H.L., Metspalu, A., Meulien, P., Miyano, S., Naparstek, Y., O'Rourke, P.P., Patrinos, G.P., Rehm, H.L., Relling, M.V., Rennert, G., Rodriguez, L.L., Roden, D.M., Shuldiner, A.R., Sinha, S., Tan, P., Ulfendahl, M., Ward, R., Williams, M.S., Wong, J.E., Green, E.D., Ginsburg, G.S., 2015. Global implementation of genomic medicine: we are not alone. Sci. Transl. Med. 7 (290), 290ps13.

Mitropoulos, K., Al Jaibeji, H., Forero, D.A., Laissue, P., Wonkam, A., Lopez-Correa, C., Mohamed, Z., Chantratita, W., Lee, M.T., Llerena, A., Brand, A., Ali, B.R., Patrinos, G.P., 2015. Success stories in genomic medicine from resource-limited countries. Hum. Genomics 9, 11.

Mitropoulos, K., Cooper, D.N., Mitropoulou, C., Agathos, S., Reichardt, J.K.V., Al-Maskari, F., Chantratita, W., Wonkam, A., Dandara, C., Katsila, T., Lopez-Correa, C., Ali, B.R., Patrinos, G.P., 2017. Genomic medicine without borders: which strategies should developing countries employ to invest in precision medicine? A new "Fast-second winner" strategy. OMICS 21 (11), 647–657.

Mlotshwa, B.C., Mwesigwa, S., Mboowa, G., Williams, L., Retshabile, G., Kekitiinwa, A., Wayengera, M., Kyobe, S., Brown, C.W., Hanchard, N.A., Mardon, G., Joloba, M., Anabwani, G., Mpoloka, S.W., 2017. The collaborative African genomics network training program: a trainee perspective on training the next generation of African scientists. Genet. Med. 19 (7), 826–833.

Nordling, L., 2017. How the genomics revolution could finally help Africa. Nature 544 (7648), 20–22.

Pang, T., 2013. Genomics for public health improvement: relevant international ethical and policy issues around genome-wide association studies and biobanks. Public Health Genomics 16 (1–2), 69–72.

Piccoli, M.L., Brito, L.F., Braccini, J., Fernando, F.F., Sargolzaei, M., Schenkel, F.S., 2017. Genomic predictions for economically important traits in Brazilian Braford and Hereford beef cattle using true and imputed genotypes. BMC Genet. 18, 2.

Sanchez, A., Faundez, P., Carvallo, P., 2011. Genomic rearrangements of the BRCA1 gene in Chilean breast cancer families: an MLPA analysis. Breast Cancer Res. Treat. 128 (3), 845–853.

Sirisena, N.D., Neththikumara, N., Wetthasinghe, K., Dissanayake, V.H.W., 2016. Implementation of genomic medicine in Sri Lanka: initial experience and challenges. Appl. Transl. Genomics 9, 33–36.

Tekola-Ayele, F., Rotimi, C.N., 2015. Translational genomics in low- and middle-income countries: opportunities and challenges. Public Health Genomics 18 (4), 242–247.

The World Bank Research and development expenditure (% of GDP) in 2015. Available from: https://data.worldbank.org/indicator/GB.XPD.RSDV.GD.ZS.

Wang, P., Chen, Z., 2013. Traditional Chinese medicine ZHENG and Omics convergence: a systems approach to post-genomics medicine in a global world. OMICS 17 (9), 451–459.

Wünsch-Filho, V., Eluf-Neto, J., Lotufo, P.A., Silva, Jr., W.A., Zago, M.A., 2006. Epidemiological studies in the information and genomics era: experience of the clinical genome of cancer project in São Paulo, Brazil. Braz. J. Med. Biol. Res. 39 (4), 545–553.

Zawati, M.H., Knoppers, B., Thorogood, A., 2014. Population biobanking and international collaboration. Pathobiology 81 (5–6), 276–285.

Zheng, S., Song, M., Wu, L., Yang, S., Shen, J., Lu, X., Du, J., Wang, W., 2010. China: public health genomics. Public Health Genomics 13 (5), 269–275.

# CHAPTER 2

# Taking Genomics From the Bench to the Bedside in Developing Countries

**Nirmala D. Sirisena, Vajira H.W. Dissanayake**
*Human Genetics Unit, Faculty of Medicine, University of Colombo, Colombo, Sri Lanka*

## BACKGROUND

Since the announcement of the completion of the Human Genome Project in 2001, genomics has continued to make a significant impact on health care. One can trace the path from genetics to disease biology as it progressed from identifying disease genes to sequencing genomes, to mapping disease genes, to cataloging common and rare genetic variations, to interpreting their clinical significance, and to understanding how the genome folds into three-dimensional maps. All these developments have significantly influenced biomedical research, leading to remarkable achievements and innovations in medical science and clinical care (Dissanayake and Barash, 2016). Although we are well into the second decade after the completion of the Human Genome Project, its full potential still remains untapped, and its clinical benefits are yet to affect the global population at large.

Advances in sequencing technologies and decreasing costs are making whole genome sequencing (WGS) and whole exome sequencing (WES) increasingly accessible and are enabling the transition from research applications and consumer genomics to routine clinical care. Translational genomics can no longer simply be considered a vague conduit from bench research to bedside care, with research only conducted in a top-down fashion involving studies done in secrecy, with little or no reporting back to research participants. Nowadays, patients contribute more to, as well as demand more from, their clinical encounters and genetic/genomic data (Isaacson Barash, 2016). Public health genomics is defined as the effective translation of genome-based knowledge and technologies for the benefit of population health. So far, the major emphasis has been on genomic "discovery" research and its impact on individual health, rather than on how such discoveries could be integrated into practice to

Genomic Medicine in Emerging Economies. http://dx.doi.org/10.1016/B978-0-12-811531-2.00002-3

evaluate its public health impact among all segments of the global population. One can argue that the emphasis should first be on cataloguing the vast range of genomic and phenotypic variations in the diverse global human population prior to developing effective interventions at the level of the individual patient (Pang, 2013). It is, however, the impact that it makes on the lives of individuals that would help the clinical implementation of genomic medicine gather momentum.

Although research in genomics and other "omics" is advancing at a rapid pace in the affluent Western world, progress has been much slower in less resourced countries, mainly because of various constraints. This inequity has raised concerns about whether the advances in genomics and the impact that it has had on health care—through the development of novel and improved diagnostics, personalized treatment, risk identification, and disease prevention—would ever be shared and become available to populations living in less resourced regions of the world (Isaacson Barash, 2016). The ability for developing nations to effectively implement genomic medicine would advance knowledge specific to their populations that is crucial to improving the health and wellbeing of the people living in those countries in the future (Helmy et al., 2016).

Implementation of genomic medicine in any country is, to a large extent, dependent on successful implementation of genome sequencing and clinical bioinformatics. This prerequisite has posed tremendous challenges for resource-constrained countries and has impeded the translation of genomic medicine from bench to bedside, creating a significant gap between them and the better resourced nations.

In 2015 in an effort to ascertain specific differences between the "haves and have-nots," the National Human Genome Research Institute and the Institute of Medicine of the National Academies of Sciences assembled 90 leaders in genomic medicine from the United States and 25 other countries across five continents to identify regional capabilities and the current state of implementation and opportunities for collaboration. An informal poll from attendees showed that most of these countries to different extents had specialized clinical genomic capabilities (e.g., cancer detection and treatment, rare disease diagnosis, and microbial pathogen identification). It is noteworthy that the poll results were similar to those in a survey done in 2012 (Isaacson Barash, 2016; Manolio et al., 2013).

## UNMET NEEDS OF THE DEVELOPING WORLD

The developing world lags behind in two areas. First, there is lack of genomic laboratory infrastructure. Second, there is a lack of a coordinated effort to impart the necessary knowledge, skills, and attitudes to the health care workforce to successfully implement genomic medicine (Isaacson Barash, 2016).

A special issue on genomic successes in the developing world in the *Journal of Applied and Translational Genomics* in 2016 served to demonstrate that the need, desire, and capacity to implement genomic medicine in developing countries exist and that pockets of excellence exist where resources and manpower are available (Dissanayake and Barash, 2016). A survey on unmet needs, conducted on participants at the Asia Pacific Society of Human Genetics Meeting in Hanoi in 2015, published in the same journal, provides insight into what is needed to achieve widespread and equitable implementation of genomic medicine (Isaacson Barash, 2016). The participants identified the lack of bioinformatics and computational tools, absence of trained data scientists, lack of access to datasets, and lack of funding as some of the reasons contributing to the disparity between the developing world and the West. Areas identified through the survey that need further attention include the following: (1) the need to convince policy makers that genomics is important and that funding genomics infrastructure and genomics education for health care workers is the crucial first step in the path to implementing genomic medicine, (2) the need for researchers to collaborate, (3) the need for different labs to share their internal data, (4) the need for global help with basic clinical research, (5) the need for affordable genetic and genomic tests, (6) the need to design laws and regulations to ensure the existence of public genomic health programs, (7) and the need to train data scientists (Isaacson Barash, 2016). Some of the research and clinical delivery needs that have become barriers to implementing genomic medicine in developing countries are illustrated in Fig. 2.1.

In view of the barriers identified above, specific measures need to be put in place to establish viable translational and precision medicine initiatives in developing countries (Isaacson Barash, 2016). These measures are summarized in Fig. 2.2.

## SRI LANKA'S INITIAL EXPERIENCE IN IMPLEMENTING GENOMIC MEDICINE

As a result of increased access to information through the Internet and high penetration of mobile technologies, news related to advances in genomics is rapidly reaching the shores of developing countries like Sri Lanka faster than ever before. In addition, expatriate communities from developing countries, such as Sri Lanka, who live in the affluent West often share information of such advances with their friends and relatives back home. In certain instances, the health care of relatives left behind in home countries is paid for by their relatives in the affluent West who demand a similar level of service in their home countries. We experience this frequently in Sri Lanka. Most of the time the local demand for genetic and genomic services that is created through such influences is unmet in developing countries such as ours because our medical

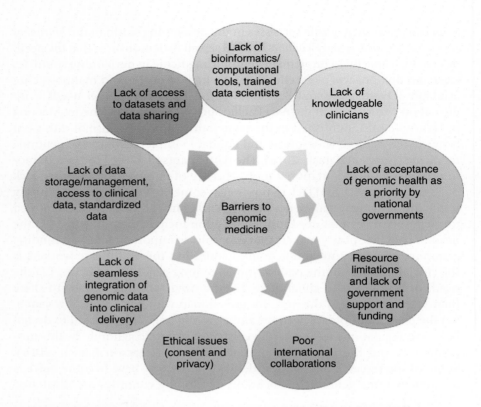

**FIGURE 2.1** Barriers to implementation of genomic medicine in developing countries.

profession is unable to provide these services. The major constraints, as stated before, are the infrastructure and manpower that enable relevant genomic tests to be performed locally (Sirisena et al., 2016).

In one of our previous papers that related several translational genetics and genomics success stories from the developing world, we described how our initial experience in implementing genomic medicine at our center in Sri Lanka led to improved patient care (Sirisena et al., 2016). To successfully integrate genomic medicine into our clinical practice, we had to improve the existing laboratory infrastructure, train our staff, and focus especially on developing our in-house bioinformatics capabilities. In addition, we had to convince technology suppliers that we were committed to implementing next-generation sequencing (NGS) services. So far we have successfully applied genomic technologies to identify genomic variants in more than 80 probands with a spectrum of genetically heterogeneous rare Mendelian disorders. Using the Illumina MiSeq NGS platform and an in-house bioinformatics pipeline established using open-source software, we successfully implemented clinical exome

Convince national governments that genomic health is important and that government support for funding a genomics infrastructure is vital

Provide adequate funding to train a genomics workforce with expertise in integrating genomic data into clinical delivery services

Provide genomic education for both clinicians and patients

Provide affordable genomic medicine services

Set up regulatory frameworks to oversee the ethical conduct of genomic research

Increase capacity of existing ethics review committees to review and monitor genomic research

Design national laws and regulations to ensure the existence of public genomic health programmes

Train data scientists and provide access to both data and genetic resources

Provide avenues for management/integration of genomic data into clinical delivery and sharing of de-identified data

Foster multidisciplinary collaborations for genomics research and services

Build public trust and confidence in genomics research, genetic data sharing, and contribution of samples and data to bio banks

**FIGURE 2.2** Measures that have to be taken to facilitate the implementation of genomic medicine in developing countries.

sequencing (CES) for rare and complex disorders with unusual coexisting phenotypes, and multigene cancer panel testing for inherited cancer syndromes. CES is performed using the TruSight One exon enrichment technology on the Illumina MiSeq NGS platform. The CES kit contains 4836 clinically relevant genes. The cancer gene panel that was implemented tests 94 genes associated with both common (e.g., breast, ovarian, endometrial, colorectal, prostate, gastric, pancreatic, and thyroid) and rare hereditary cancers. It was performed using the TruSight Cancer sequencing kit produced by Illumina, USA, on the same NGS platform.

The main challenges we encountered were the issues of who pays for the tests and how to establish the test pricing. At present the Sri Lankan National Health Service does not pay for genomic testing and there is no coverage for genomic tests by insurance companies. As such, patients have had to bear the burden of financing these tests out of their own pockets. These continue to be challenges because the cost of service contracts, reagents, other consumables, etc., are constantly escalating, and a service dependent on payment by the customer where the markup is kept deliberately small to make them affordable to those who need it cannot survive without cash infusion from philanthropic donors. Issues related to financing tests, perception about the usefulness of these tests among policymakers and administrators, and the high cost of the maintenance contract once the warranty period passed were additional challenges that we had to face (Sirisena et al., 2016). On the other hand, the challenge for bioinformatics and reporting include the interpretation of variants in the context of the phenotypic data, dealing with variants of uncertain clinical significance (VUS), nonrepresentation of variants found in the Sri Lankan population in public databases, and dealing with incidental findings. We expect that some of these challenges will be resolved as we sequence more and more exomes and understand the allele frequencies of variants in our own population. We believe that the rapid accumulation of data and the availability of updated population frequency databases will also result in improved variant classification and an overall reduction in inconclusive results (Castéra et al., 2014; Eggington et al., 2014; Hiraki et al., 2014). The fact, that we were doing our own bioinformatics analysis in-house using already well-established open-source tools and custom-written program scripts, rather than outsourcing that work as is usually the case in resource-poor settings, enabled us to have first-hand experience of handling data generated through NGS and to appreciate the quality issues related to such data, such as differences in coverage and depth in different parts of the genome. It also enabled us to have an understanding of the limitations of the commercial kits that we used and to have meaningful discussions with the manufacturers on possible requirements for customization of the gene panels for local needs. The ethical, legal, and social implications of moving from genetic to genomic counseling, managing patient privacy in an atmosphere of collaborative, open-access public data sharing, and working with families encountering legal prohibition of selective termination of pregnancies based on genetic information are some of the other challenging concerns related to genomic medicine that we have had to tackle (Sirisena et al., 2016).

The implementation of genomic medicine into our routine clinical practice has facilitated improved care for our patients (all of whom hitherto lacked a precise genetic diagnosis and appropriate treatment) by enabling us to diagnose and successfully manage patients with rare disorders, complex disorders with unusual coexisting phenotypes, and inherited cancer syndromes. It has also enabled us to build a catalog of genetic variants in the Sri Lankan population.

Experiences similar to that of our own in other developing countries, however, is often isolated, and there is therefore the need for more widespread equitable implementation of genomic medicine throughout the developing world. In our case, future plans for expanding our efforts and reaching out to a larger population include establishing public–private partnerships, lobbying for increased research funding, increasing access to research funding through international collaborations, and creating new training programs on genomic medicine aimed at health care professionals. These measures would ensure sustainable future development of genomic medicine services and research in Sri Lanka.

## OVERCOMING BARRIERS TO IMPLEMENTING GENOMIC MEDICINE IN DEVELOPING COUNTRIES

Measures to be taken to facilitate the implementation of genomic medicine in developing countries are listed in Fig. 2.2. Some of these are discussed in detail below.

- Convince national governments that genomic medicine is important and that government support for funding genomics medicine infrastructure is vital to provide affordable genomic medicine services.

The lack of access to genome sequencing facilities in developing countries makes outsourcing the only available option to utilize these technologies. Typically in developing countries, the sequencing facility of the outsourcing company is located abroad, far removed from where the output of their services is utilized. This results in many problems: improper handling of samples, improper transport of samples, long delays in generating data, quality issues related to the data generated, and interpretation and clinical correlation of results. An added danger is that the entire outsourcing process can fail at any stage, especially when the companies decide to terminate their services abruptly. The solution to this is the establishment of centers of excellence in developing countries where sequencing and bioinformatics facilities are made available. These centers can be set up at the national or regional level.

Financing the setting up of such facilities, however, may be challenging. The estimated costs start at US$100,000 for even the lowest end NGS instrument when ordered from a developing country. In fact, when the added cost of shipping, customs duties, and margins kept by local distributors are taken into account, setting up a facility can cost from 50% to 100% more in a developing country when compared with the cost in the West. In addition, in most instances other laboratory infrastructure and computational facilities also have to be established from scratch or improved. Once the facility is set up, operational costs can be prohibitive as well. Most developing countries do not have the volume of testing that the suppliers demand to bring down the price of reagents, and as a result, at the point of delivery, the cost of reagents is 50%–100% higher when compared with the cost in the West. Such huge costs are not affordable for most educational, research, and clinical institutions in developing countries. This makes establishing such facilities beyond the budget of most institutions in developing countries (Helmy et al., 2016) and thus requires investment by governments.

In that regard, making the case for investment in genomic medicine with policy makers is needed so that increased government funding can make it possible for the establishment of such centers of excellence (Penchaszadeh, 2015). In addition, governments should commit to developing well-funded and efficient systems for the evaluation and regulation of the application of genomic technologies in health care. Academic and research institutes should provide the leadership to overcome the challenges of financing by making strategic investments that promote genomics and by lobbying with governments to make budgetary provisions for the establishment of such facilities.

To reduce financial requirements, institutions and governments can consider adopting emerging low-cost technologies. Although such technologies are still in their infancy, they are nevertheless promising in many ways, offering the possibility of a wide range of genome sequencing applications that require minimal laboratory and computational skills (Helmy et al., 2016).

- Provide adequate funding to train a genomics workforce with expertise in integrating genomic data into clinical delivery services and provide genomic education for both clinicians and patients.

Genome sequencing is an interdisciplinary field that requires knowledge in molecular biology, biochemistry, and bioinformatics. In addition, when it comes to using genomic information in a clinical setting, clinicians also need to be trained in the use and interpretation of genomic information and in genomic counseling. Developing countries lag behind in all these areas (De Abrew et al., 2014). In the field of public health, public health professionals need to become knowledgeable about genes as determinants of health, how genomic testing will be used in screening, disease prevention, and health care, and its potential in the control and treatment of infectious diseases (Burton et al., 2014).

Since genomic testing is increasingly used in clinical practice (O'Daniel and Berg, 2016), efficient means of integrating genomic data into existing health records is needed, with consideration of appropriate clinical decision support tools for prompting point of care use of genomic information and delivering it to healthcare providers in an easily interpretable format. Novel biomedical informatics infrastructure and tools are essential for developing individualized patient treatment regimens based on specific genomic profiles (Madhavan et al., 2009). Implementing such strategies will offer more intensive primary or secondary prevention interventions to those at greater genomic risk. Such measures will become increasingly important in public health programs that seek to maximize benefit while minimizing adverse effects and providing cost-effective services (Burton et al., 2014). To achieve all of this, genomic and clinical data need to be integrated in a way that will allow swift and efficient translation of laboratory discoveries to the clinic. It is necessary for all health care professionals to be trained on specific core competencies regarding knowledge, skills, and attitudes in order to effectively implement genomic medicine at different levels of healthcare delivery. In this regard, new professional practice guidelines also need to be developed (De Abrew et al., 2014; Korf et al., 2014; Manolio et al., 2015). Moreover, developing countries would need to look broadly at taking the necessary steps to build a health care workforce that is adequately trained to use genomic information in their professional practice (De Abrew et al., 2014).

Education of health professionals to assess potential benefits and risks (including ethical issues) of the adequate utilization of genetic/genomic technologies in health care is also vital. As genomic sequencing is increasingly used in research with the anticipation of informing clinical health care options, a new set of decisions and dilemmas face both participants and researchers. These include how health care providers interpret and communicate results, and the ongoing need for the counseling and education of those receiving such results (Cohn et al., 2015). Developing a framework to build capacity among healthcare professionals for the use of genomics to address the health needs of the public for disease prediction, prevention, and treatment is urgently needed (De Abrew et al., 2014).

One solution would be to create a resource network to connect regional experts and related resources to researchers and data scientists in need. Online educational/training resources as well as the opportunity to connect with experts for specific research advice, knowledge transfer, collaborative problem-solving, and capacity building could well serve both researchers in training as well as translational researchers. Such networks can further benefit researchers and clinicians in resource-constrained settings by connecting to existing global initiatives designed to share approaches and lessons learned toward accelerating the implementation of genomic medicine worldwide. Examples include the Pan-Asia Pacific Genome Initiative, the Global Organization for Bioinformatics Learning, Education and Training, the Asia-Pacific Bioinformatics Network, and various centers for global health around the world (Isaacson Barash, 2016; Manolio et al., 2015).

Patient activation is defined as the skills and confidence that equip patients to become actively engaged in their own health care. Low activation is associated with unhealthy behavior, poorer outcomes, and higher costs (Fluegge, 2016). An important component necessary for the success of genomic medicine is linked to the health care consumers' understanding and acceptance of this relatively new concept. Individuals must be willing to gather and use genetic information in their health decisions, to share this information with their health care providers, and also to self-monitor and manage their health-related behaviors. Understanding how the public perceives genetics and genomics research, what their concerns and expectations are, and their attitude toward using genetic information in health decisions is critically important for the planning and provision of genetic services (Etchegary et al., 2013; Zimmern and Khoury, 2012).

- Set up regulatory frameworks to oversee the ethical conduct of genomic research.

Limited research infrastructures and poorly developed research and ethics governance mechanisms have posed challenges for both researchers and ethics committees, especially in the developing world (Dissanayake and Barash, 2016). A sustainable and equitable process from genomics research to public health interventions requires well-established ethics standards and policies, especially in developing countries. Research in this field must meet the highest ethical, legal, and socially appropriate standards, and also be accompanied by effective policies to ensure that products and outcomes target the greatest needs of public health and of individuals (Pang, 2013). For genomic medicine research to be fully translated into clinical care, it is essential for researchers to engage stakeholders who ultimately regulate the use of genomic technologies and therapeutics within health care practice (O'Daniel and Berg, 2016). Genome sequencing research and clinical applications can involve sensitive information such as personal data (e.g., name, gender, date of birth, and race), medical history, and family history of disease. Such information should be handled carefully with restricted regulations to protect the privacy and maintain the anonymity of the source of the sample, as is the case in most parts of the developed world. However, in the developing world, genome research is rare or negligible. In addition, the overall weak infrastructure and the weak or nonexistent regulatory frameworks (such as established legal and ethical conventions that stipulate quality and proficiency standards, and specimen shipment requirements) pose major hindrances that limit the capacity of a country to participate in genomic research. Often the scrutiny and rules set by international research ethics communities are not imposed, and resources for strict enforcement of these regulations are almost nonexistent. It is therefore necessary to set up regulatory frameworks to oversee the ethical conduct of genomic research as well as increase the capacity of existing ethics review committees to review and monitor genomic research. Thus research ethics committees must be put in place to guide clinicians and researchers in conducting research within the boundaries of ethical regulatory frameworks. There is also the need for development of a greater understanding of genomics among members of research ethics or scientific review committees for whom genomics may be a relatively new concept, and for whom genomics raises concerns because of lack of familiarity.

- Train data scientists and provide access to both data and genetic resources.

Bioinformaticians and data scientists are a commodity in any part of the world. The demand for such professionals will grow exponentially in the coming years. To keep abreast with this demand, education and training in the basics of genomics and bioinformatics will have to be introduced at the level of secondary education, while more advanced training can be at the undergraduate and graduate levels (Cohn et al., 2015). Access to genomic data manipulation and analysis tools is essential for providing such training. Although many bioinformatics tools are distributed under different open-source licenses, many advanced tools that are sold by companies require complicated licensing procedures to be used. Such licensure fees are frequently too expensive for institutions in the developing world.

Having access to up-to-date literature is crucial for those working in fields such as genomic medicine. The recent growth of open-access publications has greatly assisted in making contemporary literature available to scientists from developing countries. However, highly respected journals that require subscription fees are unaffordable for developing countries. This means that important new knowledge is inaccessible. In addition, genomic data download and manipulation require fast and stable Internet connections that are not always available in developing countries. Steps have to be taken to overcome these challenges.

- Provide avenues for management/integration of genomic data into clinical delivery and sharing of de-identified data.

One factor that is preventing the wider practice of genomic medicine in routine clinical practice is the limitation in our current understanding of the clinical relevance and predictive value of the genomic variants detected by NGS, and the implications of false-positive results (Adams et al., 2016; Bodian et al., 2014). The sharing of genomic data among scientists across borders is essential therefore for understanding the clinical implications of genomic variants. Reevaluation of the patchy literature regarding disease variants depends on continued data sharing and standardization of reporting (Manrai et al., 2016). It is therefore imperative to train data scientists and provide access to both genetic resources and sharing of de-identified data as well as provide avenues for genomic data management and integration into clinical delivery systems.

- Foster multidisciplinary collaborations of genomics research and services.

In the current globalized world, establishing international collaboration is the key to success in making advances in genomic medicine. In addition, collaborations among researchers, clinicians, and pharmaceutical companies in the developed and developing world are essential in conducting clinical trials that translate laboratory findings into clinically applicable therapeutics. Understanding and interpreting the molecular information gained through various laboratory techniques, such as genome sequencing and proteomics, requires that information be shared between laboratories and clinics (Goldblatt and Lee, 2010). Multidisciplinary collaborations between developed and developing nations in genomic research will undeniably result in a significant increase in capacity building for genomic research. Through such partnerships, developed nations can also provide support for establishing centers of excellence by providing funding, equipment, and training. Thus it is crucial to foster international collaborations for the advancement of genomics research and services in developing countries.

- Build public trust and confidence in genomics research and the sharing of samples and data to biobanks.

An important element in realizing the potential of genomic medicine to improve health is the growth of biobanks around the world. These are usually large repositories that contain growing numbers of individuals' genomic DNA that are linked with other health-related and lifestyle data. Biobanks are acknowledged as important resources for advancing genomics research and improving health, as they provide the keys to both basic research and facilitate translation into interventions that will ultimately lead to public health advances (Pang, 2013). Sometimes the setting up of a biobank heralds the entry of genomic medicine into a country. However, a number of participant concerns have been identified, including appropriate consent for the collection, storage and use of (medical or genetic) information, ownership and data sharing policies, and the return of individual research results (Etchegary et al., 2013). These areas of concern need to be addressed in order to foster public trust and participation in genomics services and research (Parker and Kwiatkowski, 2016).

## CONCLUSIONS

In conclusion, it is clear that there are significant barriers to overcome before genomic medicine is widely implemented in the developing world. It is well known that the World Health Organization and international development partners set the development agenda of developing countries. Genomic medicine unfortunately is not on the priority list of concerns of the WHO and other international development partners. Until and unless genomic medicine is elevated to that position, genomic medicine will continue to be ignored by governments in developing countries. The only way to overcome this situation is for those who are championing genomic medicine to work toward a World Health Assembly declaration on genomic medicine, calling on member states to implement genomic medicine in their countries. That would set the wheels in motion around the world.

## References

Adams, M.C., Evans, J.P., Henderson, G.E., Berg, J.S., 2016. The promise and peril of genomic screening in the general population. Genet. Med. 18 (6), 593–599.

Bodian, D.L., McCutcheon, J.N., Kothiyal, P., Huddleston, K.C., Iyer, R.K., Vockley, J.G., et al., 2014. Germline variation in cancer-susceptibility genes in a healthy, ancestrally diverse cohort: implications for individual genome sequencing. PLoS ONE 9 (4), e94554.

Burton, H., Jackson, C., Abubakar, I., 2014. The impact of genomics on public health practice. Br. Med. Bull. 112 (1), 37–46.

Castéra, L., Krieger, S., Rousselin, A., Legros, A., Baumann, J.-J., Bruet, O., et al., 2014. Next-generation sequencing for the diagnosis of hereditary breast and ovarian cancer using genomic capture targeting multiple candidate genes. Eur. J. Hum. Genet. 22 (11), 1305–1313.

Cohn, E.G., Husamudeen, M., Larson, E.L., Williams, J.K., 2015. Increasing participation in genomic research and biobanking through community-based capacity building. J. Genet. Couns. 24 (3), 491–502.

De Abrew, A., Dissanayake, V.H.W., Korf, B.R., 2014. Challenges in global genomics education. Appl. Transl. Genom. 3 (4), 128–129.

Dissanayake, V.H.W., Barash, C.I., 2016. Unsung heroes: genomic successes in the developing world. Appl. Transl. Genom. 9, 1–2.

Eggington, J.M., Bowles, K.R., Moyes, K., Manley, S., Esterling, L., Sizemore, S., et al., 2014. A comprehensive laboratory-based program for classification of variants of uncertain significance in hereditary cancer genes. Clin. Genet. 86 (3), 229–237.

Etchegary, H., Green, J., Dicks, E., Pullman, D., Street, C., Parfrey, P., 2013. Consulting the community: public expectations and attitudes about genetics research. Eur. J. Hum. Genet. 21 (12), 1338–1343.

Fluegge, K., 2016. The new frontier in health services research: a behavioural paradigm guided by genetics. J. Health Serv. Res. Policy.

Goldblatt, E.M., Lee, W.-H., 2010. From bench to bedside: the growing use of translational research in cancer medicine. Am. J. Transl. Res. 2 (1), 1–18.

Helmy, M., Awad, M., Mosa, K.A., 2016. Limited resources of genome sequencing in developing countries: challenges and solutions. Appl. Transl. Genom. 9, 15–19.

Hiraki, S., Rinella, E.S., Schnabel, F., Oratz, R., Ostrer, H., 2014. Cancer risk assessment using genetic panel testing: considerations for clinical application. J. Genet. Couns. 23 (4), 604–617.

Isaacson Barash, C., 2016. Translating translational medicine into global health equity: what is needed? Appl. Transl. Genom. 9, 37–39.

Korf, B.R., Berry, A.B., Limson, M., Marian, A.J., Murray, M.F., O'Rourke, P.P., et al., 2014. Framework for development of physician competencies in genomic medicine: Report of the Competencies Working Group of the Inter-Society Coordinating Committee for Physician Education in Genomics. Genet. Med. 16 (11), 804–809.

Madhavan, S., Zenklusen, J.-C., Kotliarov, Y., Sahni, H., Fine, H.A., Buetow, K., 2009. Rembrandt: helping personalized medicine become a reality through integrative translational research. Mol. Cancer Res. 7 (2), 157–167.

Manolio, T.A., Abramowicz, M., Al-Mulla, F., Anderson, W., Balling, R., Berger, A.C., et al., 2015. Global implementation of genomic medicine: we are not alone. Sci. Transl. Med. 7 (290), 290ps13.

Manolio, T.A., Chisholm, R.L., Ozenberger, B., Roden, D.M., Williams, M.S., Wilson, R., et al., 2013. Implementing genomic medicine in the clinic: the future is here. Genet. Med. 15 (4), 258–267.

Manrai, A.K., Funke, B.H., Rehm, H.L., Olesen, M.S., Maron, B.A., Szolovits, P., et al., 2016. Genetic misdiagnoses and the potential for health disparities. N. Engl. J. Med. 375 (7), 655–665.

O'Daniel, J.M., Berg, J.S., 2016. A missing link in the bench-to-bedside paradigm: engaging regulatory stakeholders in clinical genomics research. Genome Med. 8 (1), 95.

Pang, T., 2013. Genomics for public health improvement: relevant international ethical and policy issues around genome-wide association studies and biobanks. Public Health Genomics 16 (1–2), 69–72.

Parker, M., Kwiatkowski, D.P., 2016. The ethics of sustainable genomic research in Africa. Genome Biol. 17, 44.

Penchaszadeh, V.B., 2015. Ethical issues in genetics and public health in Latin America with a focus on Argentina. J. Community Genet. 6 (3), 223–230.

Sirisena, N.D., Neththikumara, N., Wetthasinghe, K., Dissanayake, V.H.W., 2016. Implementation of genomic medicine in Sri Lanka: initial experience and challenges. Appl. Transl. Genom. 9, 33–36.

Zimmern, R.L., Khoury, M.J., 2012. The impact of genomics on public health practice: the case for change. Public Health Genomics. 15 (3–4), 118–124.

# CHAPTER 3

# Genomics and Public Health: China's Perspective

**Wei Wang**

*Edith Cowan University, Perth, WA, Australia; Capital Medical University, Beijing, China; Taishan Medical University, Taian, China*

## INTRODUCTION

The official opening of the China National GeneBank (CNGB) in Shenzhen on September 22, 2016, marked a new phase in Chinese-international genomics collaboration, providing scientists from across the world with access to one of the world's most comprehensive and sophisticated biorepositories, with the goal of enabling breakthroughs in human health research and of contributing to global biodiversity conservation efforts. The CNGB is the first national gene-bank integrating a large-scale biorepository and an omics database, with the mission of collecting, preserving, and exploiting genomics resources, and of building a network fostering global communication and collaboration on biodiversity conservation and genetic resources utilization. In addition, the CNGB is supported by high-throughput sequencing and bioinformatics capacity, and it will not only provide a repository system for biological collection but, more importantly, develop a novel platform to further understand genomic mechanisms of life. The CNGB will integrate new trends in life science and international cutting-edge developments in omics to further strengthen the "three banks and two platforms" function, a model which covers the spectrum from resources to scientific research. Meanwhile, within the key fields of the biotechnology industry, such as personnalized medicine, digital healthcare, renewable energy, agriculture, and marine biology, the CNGB is now making efforts to build a platform to develop genetic informatics, and then fully promote this bioindustry innovation in China.

This reflects the great effort that China has made to improve public health through the latest technologies. On the other hand, China also has a proud history and legacy of traditional medicine that can be taken advantage of. By

Genomic Medicine in Emerging Economies. http://dx.doi.org/10.1016/B978-0-12-811531-2.00003-5

combining the latest technologies and traditional knowledge, China is able to develop a model and make a greater contribution to public health all over the world. In this chapter we review the history and current state of public health in China, China's recent developments and advancements in public health services and genomics, and the prospects of integrating new modern technologies and Traditional Chinese Medicine (TCM). China is already a major contributor in the control and management of global health burden and disease risks, as this is an inevitable aspect of China's growing international participation in the global trade of goods and services. The integration of traditional knowledge and modern technologies in genomic medicine in China will ultimately have a profound meaning for the shaping of health care and research policy while also contributing to health development of other countries, for example, in the developing world.

## THE HISTORY AND CURRENT SITUATION OF PUBLIC HEALTH IN CHINA

### Current Status of Demographics and Family Planning in China

China is a multicultural country composed of 56 ethnicities, with a diverse population of over 1.3 billion and an imbalanced economic development. The demographics of China are uneven, exaggerated by the imbalanced economy, by the large number of migrant populations, and by rapid aging. The average life expectancy is 73 years (Peng et al., 2006; Year Book, 2008a; Zhao, 1999). Compared with developed countries, China's birth rate is still high; however, there was a declining trend in the past decade. In particular, after establishing a basic national policy of family planning, the birth rate dropped from 36% in 1949 to 12.10% in 2007. Meanwhile, mortality declined from 20% in 1949 (Year Book, 1991) to 6.93% in 2007 (Year Book, 2008b). As a result, China's population has an aging trend. The United Nations predicts that more than 453 million Chinese will be older than 60 by 2050 (World, 2004).

Since 1978 the large-scale rural labor (*nongmingong* in Chinese) migration has become an eye-catching phenomenon in the process of China's social and economic development, with the deepening of reforming the open-door policy and the improvement of labor markets and related policies. The rural population has provided a vast reservoir of people willing to work for low wages in factories, at construction sites, and wherever another pair of hands is needed. Thus China's resulting exceptionally high rate of internal migration is both the consequence and the cause of economic development. Huge internal migration from rural to urban areas was estimated at 140 million in 2005—10% of the total population. Three-quarters of this migration occurred within provinces (Hu et al., 2008). Most domestic migrations are due to rural labor migrants who are almost entirely without any medical benefits in contrast to most urban residents, and they do not enjoy the state subsidies granted permanent urban residents (Wang and Zuo, 1999). They also tend to live in crowded, low-quality housing, often at the work site (Feng et al., 2002; Roberts, 1997; Shen and Huang, 2003). All these factors aggravate the dangers of disease, such as maternal and infant diseases (Asweto et al., 2016; Hu et al., 2008; Shaokang et al., 2002; Yang et al., 2005).

China established a "one family, one child" policy in 1979 and a family planning law in 2002. Under the one-child policy, couples were encouraged to marry late, usually in their mid-20s, and allowed to have only one child. This policy brought a conspicuous birth rate decline from 37.88% in 1965 to 18.21% in 1980 (Flaherty et al., 2007). In 1985 the birth rate had rebounded to 21.04% because of the "1.5 policy" of 1984, which permitted peasants whose first child was a girl to have a second child after a suitable period. Since then, the birth rate has been declining gradually. Now, it is under 15%. To date, China's one-child policy has contributed 200–400 million less people to the total growth of world population (King, 2005). The one-child policy, insofar as it limits couples to have one or two children, leads to more attention to the diagnosis of birth defects and greater involvement of parents in child care, which is named "healthy birth and child care" (Short et al., 2001). The law on maternal and infant health (launched in 1994) requires physicians to recommend a postponement of marriage if either member of a couple has a genetic disease. If one spouse has a serious hereditary disease, the couple may only marry if they agree to use long-term contraception or to undergo sterilization. If prenatal tests reveal that a fetus has a serious hereditary disease or serious deformity, the physician must advise the pregnant woman to have an abortion, and the law states that the pregnant woman is supposed to follow this recommendation. A survey on genetic research and practice was done on 402 genetic service providers in China, using a Chinese version of an internationally circulated survey questionnaire on ethics and genetics. In all, 255 participants completed the questionnaires (63%). The majority of respondents (89%) reported that they agreed with current Chinese laws and regulations on termination of pregnancy for genetic abnormalities on the basis of considerations of population control and family planning (Mao and Wertz, 1997).

## Public Health Care Structure in China

In 2005 the fund allocated to the health sector was 60,150 million *renminbi*, about 1.77% of the total government budgetary expenditure in China. In 2006 there were 308,969 health institutions in urban China and 609,128 village clinics in rural China (Health, 2007). There are wide differences in the health care systems between urban and rural areas, that is, Urban Health Care System versus Rural Health Care System, but both urban and rural health services deliver through three levels: primary, secondary, and tertiary (Shao et al., 2013).

### Urban Health Care System

In the urban system, the Center for Disease Control and Prevention (CDC) operates at provincial, city or district, and community levels (Zhao et al., 2011). Meanwhile, there are three types of hospitals in the urban medical care system, which are tertiary, secondary, and primary hospitals. A primary hospital has basic facilities and fewer than 100 beds, offering prevention, sanitation, health education, and treatment services for a specific community. The urban payment system has two stages. Before 1994 the Government Insurance Scheme and Labor Insurance Scheme were the mainstream insurances, and only the employees of government agencies, public institutions, and state-owned enterprises were covered by these two insurance schemes, which also partly covered the cost of health care for the dependants of employees. According to a national survey in nine provinces in 1986, less than 14% of the urban population was not covered by any health insurance or plan (Ministry Health, 1989). After 1994 a new urban employee basic health insurance scheme—which covered more of the urban population, including the employees of the institutions as mentioned above, foreign-invested enterprises, individual enterprises, and those who were urban inhabitants but had no stable jobs—was developed to replace the two schemes. In this scheme, government-run schemes have decreased while nonmainstream insurances (e.g., commercial schemes) have increased.

### Rural Health Care System

In the rural system there is a three-tier health care network based on an administrative relationship. The county and township tiers play a communication role between the higher and the lower tier. Village clinics deliver all health services directly to rural populations, including diagnosis, treatment, prevention and health education, vaccination, and women and children health care. Payment in the rural health care system has changed a lot since the People's Republic of China was founded in 1949. Before the economic reform was launched in 1978, over 90% of the rural population was covered by the Cooperative Medical Scheme, but from the late 1980s to 2000, when the rural economic system changed a great deal, while the Cooperative Medical Scheme was not improved, less than 10% of the rural population was covered by insurance schemes (Liu et al., 1998).

Since 2002 a new rural health care system has been established by the Chinese government and is administered by the central and territorial governments. The fund is collected from private entities, local governments, and central governments. Until 2007 there were about 685 million peasants (about 66% of the total rural population) who had health insurance. In addition, commercial health insurance (also called "private" health insurance), which is a kind of health insurance paid for by organizations other than the government, is used in both urban and rural areas. It is usually paid by the client's employer, by a union, by the client and employer sharing the cost, or by the client. Commercial health insurance can overcome the deficiencies of the rural and urban health care system in China. Everyone can choose this insurance according to his or her health or economic condition. Until now, commercial insurance schemes accounted for less than 2% of all health schemes in China.

# RECENT DEVELOPMENT OF GENOMICS-RELATED PUBLIC HEALTH IN CHINA

## Genomics-Related Public Health Programs and Services in China

Genomics is the study of the total or a part of the genetic sequence information of organisms, and attempts to understand the structure and function of these sequences and downstream biological products, which differs from genetics, the study of genes, heredity, and variation in living organisms. Genomic medicine involves clinical care that includes diagnostic, therapeutic, and the other methods for identifying and monitoring diseases by using genomic information. There is no nationwide approach to public health genomics in China. However, the opening of the State Key Laboratory of Medical Genetics in 1991 established the regulations and procedures for public health genomics-related programs and services. In addition, the National Center for Women's and Children's Health (China WCH) of the China CDC is a state-level professional organization for women's and children's health under the aegis of the China CDC (Zheng et al., 2010a). The China WCH provides the administrative hub of the China Women and Children Health Network (http://www.chinawch. com/) to facilitate coordination of relevant activities, prenatal screening, and congenital anomalies surveillance. To date—except for Qinghai, Xinjiang, Ningxia, and Tibet—22 provinces, 2 autonomous regions, and 4 municipalities in mainland China have established their own local websites/networks providing information on the laws, policies, research counseling, education, and surveillance relevant to women's and children's health care.

### Prenatal Screening

The general protocol at present for prenatal screening is predominately focused on those diseases with relatively high prevalence in China, such as Down syndrome, open neural tube defect, and also, in some cases, for trisomy 18 by using multiple serum markers. This practice varies and depends on the specific request of the doctor, generally prompted by the age of the pregnant woman, her ethnicity, or family history. Prenatal screening is usually strongly recommended to those women of advanced reproductive age or to those of advanced paternal age, since they are at an increased risk of having babies with a birth defect. A survey of 1416 new mothers on the willingness to accept prenatal screening for Down syndrome after they were provided with information on the procedure showed that 91.2% of all mothers would accept it (Qiong et al., 2008). In metropolitan territories and regions with advanced economies, almost every hospital with a department of gynecology and obstetrics can provide regular instruction for prenatal screening. Pregnant women are guided by the relevant medical staff (e.g., obstetricians) to register and set up their own healthcare documents. Thus healthcare handbooks on the prenatal period are offered to pregnant women as long as they come to the hospitals to demand a prenatal screening. Suggestions on medical termination of pregnancy will be offered by the doctor if severe fetal anomalies are detected during pregnancy and are confirmed by subsequent DNA diagnostic testing. Screening may also be performed prepregnancy. In China, the coverage rates of prenatal screening are various because of the imbalanced economic development. For example, the coverage rates in Wenzhou City and Hangzhou City in southern China, which are economically developed, were 63.15% and 60%, respectively in 2007, whereas in Gansu Province of on the western frontier area of China, which is not economically developed, the coverage rate was only 31.47% in 2006. Fees for these screening tests vary from province/hospital to province/hospital. In some advanced cities of China, such as Beijing, Tianjin, Shanghai, and Shenzhen, pregnant women with urban census registry (*Hukou*) are given the privilege of taking the special prenatal screening free of charge, and the fees are paid by their insurance or by the public medical care service. To reduce the rate of birth defects, some city governments with relatively good economies began to provide free special prenatal screening to rural pregnant women, for whom there is usually no medical insurance or free medical care provided by the publicly funded service. For example, the government of Tianjin invested over 2 million *renminbi* for such purposes in 2006. Thereby, 80% of rural pregnant women could take the privilege, and 40,000 rural newborns would benefit by this policy in Tianjin. Although the Chinese government has been providing more services on prenatal screening, not everyone can benefit from it. For example, most pregnant migrant workers are unaware of either prenatal screening or other essential health care, as they are usually in a lower social and economic class and thus disadvantaged by not being able to take part in public medical care services. Dangerous deliveries and birth defects usually happen in such communities.

### Newborn Screening

There are about 20 million births each year in China. For hyperphenylalaninemia (including phenylketonuria) alone, there are 1600–1800 new cases each year. A report (2003–04) from the National Working Committee on Children and Women under the State Council of China showed that there were 800,000–1,200,000 newborns with birth defects, about 4%–6% of all newborns. Newborn screening in China was first introduced in 1981, when screenings for phenylketonuria, congenital hypothyroidism, and galactosemia were first implemented in Shanghai. Both the law on maternal and infant health (launched in 1994) and its "action program" (launched in 2000) mandated newborn screening for congenital hypothyroidism and phenylketonuria throughout the country. Since then, variations in provisions have emerged across provinces and territories. The Guangdong and Guangxi Provinces in southern China added a glucose-6-phosphatase dehydrogenase screening program because of their local disease spectrum. Some territories in Nanjing, Wuxi, and Shanghai in southeastern China have implemented congenital adrenal cortical hyperplasia screening. In addition, some territories of Shanghai have implemented tandem mass spectrometry-based screening for amino, organic, and fatty acid metabolic disorders. The Chinese Ministry of Health has now targeted screening for hearing deficiency and piloted certain territories such as Shanghai. Until 2003 screening for hearing deficiency had covered 90% of newborns in Shanghai. In 2003 newborn screening in Beijing, Shanghai, and Guangdong covered 95% of newborns, but only 20% of newborns were screened throughout the entire country; however, the extensive countryside, small villages, and towns were omitted. A total of 133 laboratories for newborn screening was established in 2003. In China, congenital hypothyroidism and hyperphenylalaninemia (phenylketonuria) have become two essential screening items that have been implemented in all provinces and territories. To date over 30 genetic disorders, including amino acid disorders (e.g., phenylketonuria), organic acid disorders (e.g., propionic acidemia), and fatty acid oxidation disorders (e.g., carnitine transporter deficiency), can be screened in economically advanced provinces and territories by using a variety of assays, including high-performance liquid chromatography and tandem mass spectrometry. Infants diagnosed with a condition will receive a follow up appointment and families will be provided treatment and counseling by newborn screening services in each province or territory within 10–20 days after birth. Provision and fees for these screening tests vary from provinces/territories to provinces/territories as well as from hospital to hospital, and most are paid by patients as out-of-pocket expenses.

### Genetic Testing

In China several laboratories for genetic testing have been founded. The genetic tests mainly deal with diseases such as breast cancer, carcinoma of the large intestine, heart rate abnormality, and nerve/muscle defects, which are taken as a kind of auxiliary examination for early prevention, early diagnosis, and early treatment. The State Key Lab of Medical Genetics of China was founded in 1984 in Hunan Province and opened in 1991. It is the only designated organization to authenticate the identification of novel abnormal chromosomal karyotypes in China. This laboratory has been maintaining the database of the family collection of genetic diseases in China and has been collecting novel human abnormal chromosomal karyotypes identified in China for the past 30 years. The database assembled 2144 novel human abnormal karyotypes identified in China (excluding those from Tibet, Taiwan, Hong Kong, and Macau) and their related disease information. The karyotypes include 4510 chromosome breakages and 57 balanced translocations between the X chromosome and autosomes. The database also provides statistical information, chromosome breakage frequencies, abnormal karyotype descriptions, and related clinical information. Some tests are paid by insurance in advanced cities (e.g., test on human papilloma virus/HPV in Beijing), but, again, most test are paid by patients as out-of-pocket expenses.

## Projects Related With the Human Genome in China

In past decades China has proposed or entered several projects related to the human genome. The earliest project started in 1991 when the Human Genome Diversity Project was generated. The Chinese Human Genome Diversity Project has collected cell lines from the 56 Chinese ethnic groups and tested the DNA samples. In 1999 China entered the International Human Genome Project and undertook the sequencing of 1% of the human genome working draft (i.e., the region 3pter-D3S3610 containing 30 million bp). China is also responsible for about 10% of the International HapMap Project, which was launched in 2002 jointly by the United States, the United Kingdom, Japan, China, and Canada. This project aimed to guide the design and analysis of genetic association studies, shed light on structural variation and recombination, and identify loci that have been subjected to natural selection during human evolution. In January 2008 the "Yanhuang" (Emperor Yan) project was established by the Beijing Genomics Institute, which aimed to sequence the entire genome of 100 Chinese individuals over a 3-year period.

## Genetics Education

Genetics education in China started fairly late because of the underdeveloped education policies of the Cultural Revolution (1966–76). In the 1980s colleges began to offer genetics courses, but only the basics of human genetics were taught in regular colleges and universities. Now, the main form of Chinese genetics education for the public is genetic counseling in clinics. Since the law on maternal and infant health was passed in 1994, genetic counseling in clinics has been developed in advanced hospitals, mainly as a unit attached to the department of gynecology and obstetrics. However, compared with developed countries, there are still many problems associated with genetic counseling in China (Zhao et al., 2014). First, genetic counseling clinics are almost all located in big cities, whereas over 50% of the population live in rural areas where people have little medical knowledge but more children Second, genetic counseling in China is now not offered by professionally trained genetic counselors but by clinicians such as pediatricians or obstetricians (Ren et al., 2002). Third, there is no official genetic counseling program in China (Zhang and Zhong, 2006). In 2007 the first website, named China Genetic Counseling Network, on genetic counseling and genetic education was established in China (http://www.gcnet.org.cn). It introduced the basic knowledge of genetic diseases, their clinical symptoms, diagnosis, preventions, and treatments.

## NEW PROSPECTS—WHEN MODERN TECHNOLOGIES MEET TRADITIONAL SKILLS

### The Concepts of Genomics and Traditional Chinese Medicine

TCM consists of a broad range of medical practices sharing common concepts and an integrated theoretical system that has evolved through over 3000 years of clinical and pharmacological trails; it is one of the oldest medical traditions in China and the Asia-Pacific region (Yu et al., 2006). As the second most practiced medical system in the world, one-quarter of the world's population uses one or more TCM therapies, including acupuncture, moxibustion, Chinese herbal medicine, Tui-Na massage, dietary therapy and physical exercise (Tai Chi and Qi Gong) to maintain health and wellness. In China, TCM shares equal status and has been integrated with Western medicine in the health care system to diagnose, prevent, and treat many types of diseases. Meanwhile, compared with Western medicine, the TCM approach has not been recognized internationally because of a lack of systemic research and investigation showing evidence of its effectiveness (Yun et al., 2012a). Nevertheless, scientists have begun to provide novel insights into the essence and molecular basis of TCM, with the globalization of life sciences (global health) and arrival of mega data mining (big data) as well as the progress of genomics research (Wang and Chen, 2013). The fast-emerging integration of the science of genomics and the practice of TCM may provide an unprecedented opportunity to improve the health of mankind (Wang, 2016; Yun et al., 2012b).

As an example of the rapidly expanding knowledge of the human genome to clinical practice, genomic medicine plays an important role in the modernization of TCM by combining TCM theory with modern biological and genomic/genetic concepts, for example, by elucidating the active components of TCM herbal medications and their pharmacodynamic mechanisms at molecular level. Personalized medicine, an approach that emphasizes the customization of health care, is used to help tailor interventions to maximize health outcomes, and is rapidly becoming a reality for many patients suffering from a variety of conditions.

The study of genomics has led to insights on gene regulation and the complex interplay of factors at play in both Chinese medicine and in personalized medicine (Yun et al., 2012a). Clinicians and life scientists, therefore, are currently at a critical junction to accelerate both TCM and its evidence base with the availability of genomics, as well as postgenomics technologies such as functional genomics, proteomics, metabolomics, glycomics, and lipidomics at molecular and cellular levels. This chapter provides an outlook on the enormous promise anticipated from the integration of TCM with genomics/genetics as a new driver for novel molecular-targeted personalized medicine, and the future directions and challenges in this hitherto neglected dimension of postgenomic-personalized medicine. Recently, this has mainly involved the systematic use of patients' genotypes and clinical phenotypes to optimize the individual's preventive and therapeutic care (Wang and Chen, 2013; Yun et al., 2012b).

## Application of Genomics Technologies in the Authentication of Traditional Chinese Medicine

The elements of TCM diagnoses are influenced by three principal factors: heredity (genetic composition), ontogeny (stage of development), and environment (climate, associated flora, soil, and method of cultivation). Genetic analyses of Chinese herbs can provide leads for the botanical identity of TCM constituents as the DNA profiling of a herbal species does not vary with their physical form, physiology, and external conditions. Natural products are gaining increased applications in drug discovery and development. DNA mutation and polymorphism analysis lead to the applications of DNA microarrays in pharmacodynamics, pharmacogenomics, toxicogenomics, and quality control of herbal drugs and extracts. Genomic analyses of Chinese herbs provide the botanical identity of TCM constituents (Hon et al., 2003). One of the most well-studied herbs is ginseng, which has long been used to maintain physical vitality in China and the Far East. Ginsenosides, a main element of ginseng, can inhibit early antigen activation of Epstein-Barr virus and also shows anticarcinogenic effects in a two-stage mouse skin model with 9,10-dimethyl-1,2-benzanthracene and in lung carcinogenesis induced by 4-nitroquinolin-1-oxide (Konoshima et al., 1996; Zhang et al., 2013). The DNA probe method for the identification of host-specific DNA fragments has been employed in DNA fingerprinting analysis of ginseng and generated a distinctive banding pattern, with a homologous index of 0.55 between Chinese and American ginseng (Ho and Leung, 2002). In another investigation, the random amplified polymorphic DNA (RAPD) technique has been used to identify the *Panax* species and their adulterant, and distinct RAPD fingerprints of American and Chinese ginseng have been obtained, irrespective of sources and ages (Shaw and But, 1995). Restriction fragment length polymorphic DNA is also applied on ginseng authentication based on the ribosomal ITS1-5.8S-ITS2 region and the 18S rRNA gene and has shown promising results on authentication of American and Chinese *Panax* species (Fushimi et al., 1997; Ngan et al., 1999).

Most Chinese herbal formulas consist of several individual herbal components, which are obviously more complicated than individual herbs. Genetic technologies have also been used to reveal the relationship between formula and components, for example, in research of the herbal formula of *San Huang Xie Xin Tang* (SHXXT); a genomics microarray 308 was used to analyze the putative mechanism of SHXXT and to define the relationship between SHXXT and its individual herbal components. Gene expression profiles of HepG2 cells treated with SHXXT's components were obtained by DNA microarray, indicating that SHXXT's components display a unique antiproliferation pattern via the p53 protein through DNA damage signaling pathways in HepG2 cells. In addition, hierarchical clustering analysis has shown that Rhizoma Coptis, the principle herb, shares a similar gene expression profile with SHXXT (Cheng et al., 2008). These findings may explain why Rhizoma Coptis exerts a major effect in the herbal formula of SHXXT. This is a good example that reveals the relationship between formula and herb (Ngan et al., 1999).

## Elucidation of Putative Biological Mechanisms of Traditional Chinese Medicine by Genomic Approaches

In recent years, genomic and molecular approaches have been extensively used to illustrate potential mechanisms and biological functions of TCM. As mentioned earlier, genomic approaches were used to authenticate ginseng from different countries. In addition, based on DNA microarray analysis, ginseng was reported to up-regulate the expression of a set of genes involved in adhesion, migration, and cytoskeleton (Yue et al., 2007).

Berberine, a well-known component of the Chinese herb medicine of *Huanglian* (*Coptis chinensis*), is capable of inhibiting growth and endogenous platelet-derived growth factor (PDGF) synthesis in vascular smooth muscle cells after in vitro mechanical injury. It also acts on suppressing PDGF-stimulated cyclin D1/D3 and cyclindependent kinase (Cdk) gene expression. Moreover, berberine has increased the activity of AMP-activated protein kinase, which leads to phosphorylation of p53 and increased protein levels of the Cdk inhibitor p21Cip1. These observations offer a molecular explanation for the antiproliferative and antimigratory properties of berberine (Liang et al., 2008).

Numerous Chinese herbs have been suggested to have antitumor potential. Scientists have found that Chinese herbs have shed light on possible mechanisms and provided biological clues for the development of new modern drugs. Konkimalla et al. (2008) showed that cytotoxicity of its derivative, artesunate, is associated with inhibition of inducible nitric oxide syntheses (iNOS). That a number of genes are involved in nitric oxide (NO) signaling and are significantly up- or downregulated by artesunate indicates that artesunate may not only inhibit iNOS, but also affect other NO-related genes. Microarray analysis also showed that the Wnt/β-catenin signaling pathway, which plays an important role in colon cancer etiology, is regulated by artesunate, and colon cancer cell lines are the most sensitive to artesunate among all solid tumor cell lines. These results collectively suggest that artesunate might attenuate the growth of human colorectal carcinoma by inhibition of the Wnt/β-catenin pathway (Konkimalla et al., 2008).

Nuclear factor-κB (NF-κB) is critically important for tumor cell survival, growth, angiogenesis, and metastasis. One of the key events in the NF-κB signaling is the activation of the inhibitor of NF-κB kinase (IKK) in response to stimuli of various cytokines. The root of *Euphorbia fischeriana* Steud. has been used as a traditional Chinese herb for more than 2000 years. The compound 17-acetoxyjolkinolide B (17-AJB), one of the components of *E. fischeriana* Steud., is a novel small molecule inhibitor of IKK. Indeed, 17-AJB has been shown to effectively inhibit tumor necrosis factor-A-induced NF-κB activated ion and induce apoptosis of tumor cells. Detailed analysis revealed that 17-AJB keeps IKK in its phosphorylated form irreversibly to inactivate its kinase activity, leading to its failure to activate NF-κB. The effect of 17-AJB on IKK is specific and has no effect on other kinases such as p38, p44/42, and JNK. The effects of 17-AJB on apoptosis also correlate with inhibitions of expressions of the NF-κB-regulated genes. It is suggested that 17-AJB is a novel type NF-κB pathway inhibitor, and its unique interaction mechanism with IKK may render it a strong apoptosis inducer of tumor cells and a novel type anticancer drug candidate (Yan et al., 2008).

Certain types of mushrooms are also considered as Chinese herbs and have widely been used as dietary supplements in the United States. Recent studies demonstrated that mushroom intake could protect against cancer, which might be linked to the modulation of the immune system (Wasser, 2002). Using DNA microarray analysis, Jiang and Sliva (2010) have found that mushroom could inhibit expression of genes involved in cell cycle regulation, thus inhibiting the invasiveness of breast cancer.

It is interesting to note that Jing (Meridian) theory is one of the foundational principles of TCM where the concept of Jing plays a central role. Presumably, the Jing concept refers to the genetic information as well as to its plasticity, as Jing is thought to be "the substance essential for development, growth, and maturation," and "conception is made possible by the power of Jing, growth to maturity is the blossoming of Jing, and the decline into old age reflects the weakening of the Jing" (Hu et al., 2012). With the availability of functional genomics and proteomics, it is now possible to examine the precise molecular targets impacted by TCM preparations and related health interventions. Ultimately, this can lead to a more evidence-informed practice of TCM that can in the near future account for individual and population variability in these molecular TCM targets. Similarly, genome-wide association studies (GWAS) can offer molecular leads on the molecular genetic substrates of the TCM mechanism of action.

## Application of Genomics Theory to Support Acupuncture Practice

Acupuncture utilizes fine needles to pierce through specific anatomical points (positioned "Jing"); Acupuncture has been extensively used and has emerged as an important modality of complementary and alternative therapy to Western medicine (Yan et al., 2013). Systems biology has become practically available and resembles acupuncture in many aspects, and is a key technology that serves as the major driving force for the translation of acupuncture medicine into practice, and will advance acupuncture therapy into health care for individuals. High-throughput genomics, proteomics, and metabolomics in the context of systems biology have been able to identify potential candidates for the effects of acupuncture and provide valuable information toward understanding mechanisms of the therapy. To realize the full potential of TCM acupuncture, the current status of principles and practice of acupuncture should be integrated with the systems biology platform in the postgenomic era (Zheng et al., 2010b). Some characteristic examples are presented to highlight the application of this platform in omics and systems biology approaches to acupuncture research as well as some of the necessary milestones for moving acupuncture into mainstream health care (Zhang et al., 2013).

## Application of Genomic Medicine to Investigate Herb–Drug Interactions

In view of the increasing use of herbal medicines not only in China and the Asia-Pacific region but also in many other parts of the world, including Western countries, concerns have been raised about herb–drug interactions. Pharmacogenomic studies are required for a better understanding of the genomic components of kinetic and dynamic effects of TCM preparations and their physiochemical active ingredients. Similarly, more studies are needed for the role of genetics for herb–drug and TCM-drug interactions (Hu et al., 2012). Accumulating evidence has demonstrated that concurrent administration of herbal remedies may alter the pharmacokinetic or pharmacodynamic behaviors of certain drugs, and thus severe adverse effects may occur (Fugh-Berman, 2000). However, mechanisms underlying herb–drug interactions remain an understudied area of pharmacokinetic and pharmacotherapy. Systematic evaluation of herbal product–drug interaction liability, as is routine for new drugs under development, necessitates identifying individual constituents from herbal products and characterizing the interaction potential of such constituents (Brantley et al., 2013). Genomic approaches applied in herb–drug interaction research should gradually illuminate such interactions and how they may be influenced by multiple environmental and/or genetic factors.

Tian Xian is a traditional Chinese herbal anticancer remedy that activates human pregnane X receptor (PXR) in cell-based reporter gene assays. Tian Xian products are herbal dietary supplements manufactured in China that are distributed worldwide and aggressively marketed as anticancer herbal therapies. These products are also marketed as herbal therapies that alleviate the unpleasant side effects associated with the anticancer treatments of Western medicine. Activation of PXR in liver regulates the expression of genes encoding proteins that are intimately involved in the hepatic uptake, metabolism, and elimination of toxic compounds from the body. PXR-mediated herb–drug interactions can have undesirable effects in patients receiving combination therapy. Tian Xian can alter the strength of interaction between the human PXR protein and transcriptional cofactor proteins. It can increase expression of Cyp3a11 in primary cultures of rodent hepatocytes and induce expression of CYP3a4 in primary cultures of human hepatocytes. These data indicate that coadministration of Tian Xian is probably contraindicated in patients undergoing anticancer therapy with conventional chemotherapeutic agents (Lichti-Kaiser and Staudinger, 2008).

Another example is the recent report on the interaction of PHY906, a TCM for chemotherapy-associated side effects relief, with Irinotecan (CPT-11), a drug used for the treatment of cancer using a systems biology approach. A study demonstrated that PHY906 significantly amplifies the effects of CPT-11 in tumor tissues. Furthermore, administration of PHY906 together with CPT-11 could trigger unique changes that are not triggered by either alone, suggesting the enhanced antitumor activity of the combination (Wang et al., 2011). Thus genomic medicine may provide a crucial link in our understanding of the complicated interactive relationships between TCM and Western drugs.

## TRADITIONAL CHINESE MEDICINE AND NEW CONCEPTS OF PREDICTIVE, PREVENTIVE AND PERSONALIZED MEDICINE IN DIAGNOSIS AND TREATMENT OF SUBOPTIMAL HEALTH

Suboptimal health status (SHS) is a physical state between health and disease, characterized by the perception of health complaints, general weakness, chronic fatigue, and low energy levels (Yan et al., 2009). SHS is foreshadowed by the an ancient concept in TCM similar to the perspective of predictive, preventive, and personalized (precision) medicine (PPPM) (Wang et al., 2016a,b,c). It is also found that SHS is associated with cardiovascular risk factors and may contribute to the development of cardiovascular disease. SHS has also been reported to be associated with chronic psychosocial stress (Kupaev et al., 2016; Yan et al., 2012, 2015) and poor lifestyle factors (Bi et al., 2014; Chen et al., 2014). Since ancient times, TCM has identified a physical status between health and disease (Wang and Yan, 2012; Wasser, 2002; Yan et al., 2008).

The existence of a reliable and valid instrument to assess SHS is essential. Wang et al. (2014) developed a tool, the SHS questionnaire-25 (SHSQ-25), to assess five components of health (Wang et al., 2014; Yan et al., 2009). The SHSQ-25 accounts for the multidimensionality of SHS by encompassing the following domains: fatigue (Zhao, 1999), the cardiovascular system (Year Book, 2008a), the digestive tract (Peng et al., 2006), the immune system (Year Book, 1991), and mental status (Year Book, 2008b). The SHSQ-25 is short and easy to comprehend and, therefore, an instrument suitable for use in both large-scale studies of the general population and routine health surveys (Wang et al., 2014). The validity and reliability of this approach were evaluated in a small pilot study and then in a cross-sectional study of 3405 participants in China. A correlation between SHS and systolic blood pressure, diastolic blood pressure, plasma glucose, total cholesterol, and high-density lipoprotein (HDL) cholesterol among men, and a correlation between SHS and systolic blood pressure, diastolic blood pressure, total cholesterol, triglycerides, and HDL cholesterol among women were detected. An ongoing longitudinal SHS cohort survey (China Suboptimal Health Cohort Study, COACS) consisting of 50,000 participants will provide a powerful health trial for the use of SHSQ-25 and its application to PPPM through patient stratification and therapy monitoring by using innovative technologies of predictive diagnostics and prognosis. To date the SHSQ-25 as a self-reported survey tool has been validated in various populations, including European ethic groups (Kupaev et al., 2016; Yan et al., 2015; Wang et al., 2017), and currently the SHSQ-25 has also been applied to a real-life community-based health survey in Ghana, Africa (Adua et al., 2017). SHS thus has been recognized internationally and is used as a novel tool for the early detection of chronic disease (Kupaev et al., 2016; Yan et al., 2015).

The availability of reliable biomarkers for noncommunicable chronic diseases (NCD) is essential for improving early detection and intervention. Specific biomarkers, such as plasma glycome or serum peptidome, are believed to represent an "intermediate phenotype" in the etiology of adult-onset diseases (Lu et al., 2011; McLachlan et al., 2016; Wang et al., 2016a,b). Therefore these profiles might hold the key to understanding the underlying biological mechanisms that create SHS. SHSQ-25 affords a window of opportunity for early detection and intervention, contributing to the reduction of chronic disease burdens. The inclusion of the "objective" biomarkers and the subjective "SHS" assessment into population studies is therefore believed to be timely in improving chronic disease control and in strengthening opportunities for chronic disease prevention.

For both developing and developed countries, the integrative concept of PPPM would enable clinicians and public health workers to predict an individual's predisposition in order to provide targeted preventive measures before the actual onset of disease. By using innovative PPPM tools—such as the SHSQ25—biomarkers (medical imaging or pathology-specific molecular patterns, sub/cellular imaging, and omics), pharmacogenetics, disease and patient modeling, individual patient profiles, and a combination of Western medicine and TCM, the expected outcomes are conducive to more effective population screening, prevention measures early in life, identification of persons who are at risk, stratification of patients for optimal therapy planning and prediction, and reduction of adverse drug-drug or drug-disease interactions. With the support of the rapid progress of biotechniques and the availability of large health databases such as the Human Genome Project, health professionals are in a good position to address the topics of genetics, environment, and behavior and to motivate the introduction of PPPM into daily medical services.

## Challenges and Opportunities

Our understanding of TCM has advanced greatly with data-intensive genomics and omics biotechnologies. However, a number of critical challenges remains ahead to fully articulate this vision and roadmap. First, it is difficult to combine multidimensional omics data emerging from epigenomics, proteomics, metabolomics, and clinical phenotype data in relation to TCM health outcomes. Second, there are complicated exogenous components of TCM. Potential personalized medicine could be found if we can utilize genomics and metabolomics techniques for full spectrum of metabolites of TCM. Genomics, together with metabolomics, transcriptomics, and proteomics, jointly inform the "systems biology" approaches that are crucial for the integration of the postgenomics knowledgebase with TCM.

Next-generation sequencing (NGS) and multi-omics integrative biology research offer new opportunities into the way we research and understand an illness like stroke. For example, the exploration of possible genetic factors in the development of complex multigeneic ischemic stroke (IS) was first made feasible by the advent of GWAS. One study utilized genomic data from a biobank (Jiang et al., 2006) to explore IS stroke prevalence in China and found a strong genetic predisposition to IS. The ability to process pedigree data could also enable the use of classical linkage techniques for the analysis of NGS in complex diseases such as IS. However, the type and amount of NGS data has been increasing rapidly. To be able to store and analyze this increasing amount of data, extremely high-performance computing and intensive bioinformatics support must be available (Zhao et al., 2012). Therefore, the opening of the CNGB, which is supported by the world's most advanced and sophisticated high-throughput sequencing and bioinformatics capacity, will provide an unprecedented opportunity to conquer diseases through genomic tools.

However, integrating TCM with modern medicine at the omics level in order to contribute to more targeted personalized medicine is a long and arduous process that requires the accumulation and synthesis of knowledge in many fields, including genomics/genetics, molecular biology, pharmacology, epidemiologic studies on gene-disease associations, gene-environment-behavior interactions, and genomics-informed clinical trials of TCM health interventions. Advances in omics may also provide new opportunities for TCM in public health that focus on disease prevention, because genomic studies can provide genetic information at the individual level as well as at the population level, thus increasing the interaction and the interdependence between the traditional healthcare delivery system, which focuses on treatment of individuals, and the public health system, which focuses on prevention, suboptimal health management, and control in populations (Wang et al., 2014). Notably, this enhanced interaction is creating a shared population health information focus on using genomic advances appropriately and effectively to promote health and to prevent disease, which can be applied to TCM its prevention aspect. Ultimately, with increasing integration, the modern form of TCM will continue to progress in the postgenomics era for a more integrated, evidence-based, and personalized health care. Hence TCM has the potential to be a new and rapidly emerging driver for novel molecular-targeted personalized medicine in the coming decade.

## CONCLUSIONS AND OUTLOOK

China has a large population spread over a large land mass; in addition, China suffers from imbalanced economic development. Although China has invested in the basic research of genome science, it is urgent for China to develop public health genomic programs and clinical genetic services that meet the needs of the Chinese people, particularly residents in rural areas. A standard framework for evaluating genetic/genomic testing and related interventions in China is also needed. What is more, the Chinese government should implement genomic education into the Chinese health care system. China needs an easy way for the public to obtain genetic information on innovation, education, and prevention. The understanding of TCM has advanced greatly along with data-intensive genomics and omics biotechnologies. However, integrating TCM with modern medicine at the omics level in order to contribute to a more targeted personalized medicine is a long and arduous process that requires accumulation and synthesis of knowledge in many fields. Advances in omics will provide new opportunities for TCM in public health that focus on disease prevention, because genomic studies can provide genetic information at individual level as well as in the population as a whole, thus increasing the interaction and the interdependence between the traditional healthcare delivery system, which focuses on treatment of individuals, and the public health system, which focuses on prevention, suboptimal health management, and control in populations. China is already a major contributor in the control and management of the global health burden and disease risks, as this is an inevitable aspect of China's growing international participation in the global trade of goods and services. The integration of traditional knowledge and modern technologies in genomic medicine in China will ultimately have profound meaning for the shaping of health care and research policy, while also contributing to the development of health care in other countries, for instance, in those of the developing world (Forero et al., 2016).

## Acknowledgments

Author acknowledge that parts of the data presented in this chapter have previously been published in our earlier articles (Adua et al., 2017; Asweto et al., 2016; Kupaev et al., 2016; Lu et al., 2011; McLachlan et al., 2016; Peng et al., 2006; Shao et al., 2013; Wang, 2016; Wang and Yan, 2012; Wang et al., 2014, 2016a–c; Wang et al., 2017; Yan et al., 2009, 2012, 2013, 2015; Yun et al., 2012a,b; Zhao et al., 2011, 2014; Zheng et al., 2010a,b). This chapter uses the original data but provides a new interpretation based on the innovative paradigm of predictive, preventive and personalized medicine.

## References

Adua, E., Roberts, P., Wang, W., Wang, et al., 2017. Incorporation of suboptimal health status as a potential risk assessment for type II diabetes mellitus: a case-control study in a Ghanaian population. EPMA J. 8, 345–355.

Asweto, C.O., Alzain, M.A., Andrea, S., Alexander, R., Wang, W., 2016. Integration of community health workers into health systems in developing countries: opportunities and challenges. Fam. Med. Community Health 4 (1), 37–45.

Bi, J., Huang, Y., Xiao, Y., Cheng, J., Li, F., Wang, T., et al., 2014. Association of lifestyle factors and suboptimal health status: a cross-sectional study of Chinese students. BMJ Open 4 (6), e005156.

Brantley, S., Argikar, A., Lin, Y.S., Nagar, S., Paine, M.F., 2013. Herb-drug interactions: challenges and opportunities for improved predictions. Drug Metab. Dispos.

Chen, J., Cheng, J., Liu, Y., Tang, Y., Sun, X., Wang, T., et al., 2014. Associations between breakfast eating habits and health-promoting lifestyle, suboptimal health status in Southern China: a population based, cross sectional study. J. Transl. Med. 12 (1), 348.

Cheng, W.Y., Wu, S.L., Hsiang, C.Y., Li, C.C., Lai, T.Y., Lo, H.Y., et al., 2008. Relationship between San-Huang-Xie-Xin-Tang and its herbal components on the gene expression profiles in HepG2 cells. Am. J. Chin. Med. 36 (04), 783–797.

Feng, W., Zuo, X., Ruan, D., 2002. Rural migrants in Shanghai: living under the shadow of socialism. Intl. Migr. Rev. 36 (2), 520–545.

Flaherty, J.H., Liu, M.L., Ding, L., Dong, B., Ding, Q., Li, X., et al., 2007. China: the aging giant. J. Am. Geriatr. Soc. 55 (8), 1295–1300.

Forero, D.A., Wonkam, A., Wang, W., Laissue, P., Lopez-Correa, C., et al., 2016. J. Med. Genet. 53 (7), 438–440.

Fugh-Berman, A., 2000. Herb-drug interactions. Lancet 355 (9198), 134–138.

Fushimi, H., Komatsu, K., Isobe, M., NAMBA, T., 1997. Application of PCR-RFLP and MASA analyses on 18S ribosomal RNA gene sequence for the identification of three ginseng drugs. Biol. Pharm. Bull. 20 (7), 765–769.

Health Statistical Year Book, 2007. Beijing: Ministry of Health; 2007. Available from: www.moh.gov.cn/publicfiles/business/htmlfiles/zwgkzt/ptjnj/200807/37168.htm.

Ho, I., Leung, F., 2002. Isolation and characterization of repetitive DNA sequences from *Panax ginseng*. Mol. Genet. Genomics 266 (6), 951–961.

Hon, C.C., Chow, Y.C., Zeng, F.Y., Leung, F.C., 2003. Genetic authentication of ginseng and other traditional Chinese medicine. Acta Pharm. Sin. 24 (9), 841–846.

Hu, X., Cook, S., Salazar, M.A., 2008. Internal migration and health in China. Lancet 372 (9651), 1717–1719.

Hu, M., Wang, D.Q., Xiao, Y.J., Mak, V., Tomlinson, W.L.B., 2012. Herb-drug interactions: methods to identify potential influence of genetic variations in genes encoding drug metabolizing enzymes and drug transporters. Curr. Pharm. Biotechnol. 13 (9), 1718–1730.

Jiang, J., Sliva, D., 2010. Novel medicinal mushroom blend suppresses growth and invasiveness of human breast cancer cells. Intl. J. Oncol. 37 (6), 1529.

Jiang, B., Wang, W.Z., Chen, H., Hong, Z., Yang, Q.D., Wu, S.P., et al., 2006. Incidence and trends of stroke and its subtypes in China. Stroke 37 (1), 63–65.

King, M., 2005. China's infamous one-child policy. Lancet 365 (9455), 215–216.

Konkimalla, V.B., Blunder, M., Korn, B., Soomro, S.A., Jansen, H., Chang, W., et al., 2008. Effect of artemisinins and other endoperoxides on nitric oxide-related signaling pathway in RAW 264.7 mouse macrophage cells. Nitric Oxide 19 (2), 184–191.

Konoshima, T., Takasaki, M., Tokuda, H., Masuda, K., Yoko, A.R., Shiojima, K., et al., 1996. Anti-tumor-promoting activities of triterpenoids from ferns. I. Biol. Pharm. Bull. 19 (7), 962–965.

Kupaev, V., Borisov, O., Marutina, E., Yan, Y.X., Wang, W., 2016. Integration of suboptimal health status and endothelial dysfunction as a new aspect for risk evaluation of cardiovascular disease. EPMA J. 7 (1), 19.

Liang, K.W., Yin, S.C., Ting, C.T., Lin, S.J., Hsueh, C.M., Chen, C.Y., et al., 2008. Berberine inhibits platelet-derived growth factor-induced growth and migration partly through an AMPK-dependent pathway in vascular smooth muscle cells. Eur. J. Pharmacol. 590 (1), 343–354.

Lichti-Kaiser, K., Staudinger, J.L., 2008. The traditional Chinese herbal remedy Tian Xian activates pregnane X receptor and induces CYP3A gene expression in hepatocytes. Drug Metab. Dispos. 36 (8), 1538–1545.

Liu, Y., Rao, K., Fei, J., 1998. Economic transition and health transition: comparing China and Russia. Health Policy 44 (2), 103–122.

Lu, J.P., Knezevic, A., Wang, Y.X., Rudan, I., Campbell, H., Zou, Z.K., et al., 2011. Screening novel biomarkers for metabolic syndrome by profiling human plasma N-glycans in Chinese Han and Croatian populations. J. Proteome Res. 10 (11), 4959–4969.

Mao, X., Wertz, D.C., 1997. China's genetic services providers' attitudes towards several ethical issues: a cross-cultural survey. Clin. Genet. 52 (2), 100–109.

McLachlan, F., Timofeeva, M., Bermingham, M., Wild, S., Rudan, I., 2016. A case-control study in an Orcadian population investigating the relationship between human plasma N-glycans and metabolic syndrome. J. Glycomics Lipidomics 6 (139), 2153–2637.

Ministry of Health. Urban Medical Service Study in China – Survey Data on Medical Care Demand and Utilization from Nine Provinces and Municipalities. Beijing: Ministry of Health; 1989.

Ngan, F., Shaw, P., But, P., Wang, J., 1999. Molecular authentication of Panax species. Phytochemistry 50 (5), 787–791.

Peng, X., Song, S., Sullivan, S., Qiu, J., Wang, W., 2006. Ageing, the urban-rural gap and disability trends: 19 years of experience in China-1987 to 2006. PLoS One 5 (8), e12129.

Qiong, E., Tang, Z.L., Huang, J.Y., Han, P., Tian, D., Li, C.L., et al., 2008. An analysis on will survey of prenatal screening for Down syndrome. Chin. Prim. Health Care 8, 33–35.

Ren, A.G., Wang, L.N., Zhao, P., Li, Z., 2002. Current status of genetic counselling in maternal and child health care institutions. Chin. J. Reprod. Health 13 (3), 131–134.

Roberts, K.D., 1997. China's "tidal wave" of migrant labor: what can we learn from Mexican undocumented migration to the United States? Intl. Migr. Rev., 249–293.

Shao, S., Zhao, F., Wang, J., Feng, L., Lu, X., Du, J., et al., 2013. The ecology of medical care in Beijing. PLoS One 8 (12), e82446.

Shaokang, Z., Zhenwei, S., Blas, E., 2002. Economic transition and maternal health care for internal migrants in Shanghai, China. Health Pol. Plann. 17 (Suppl. 1), 47–55.

Shaw, P.C., But, P.P., 1995. Authentication of Panax species and their adulterants by random-primed polymerase chain reaction. Planta Med. 61 (05), 466–469.

Shen, J., Huang, Y., 2003. The working and living space of the 'floating population' in China. Asia Pac. Viewpoint 44 (1), 51–62.

Short, S.E., Zhai, F., Xu, S., Yang, M., 2001. China's one-child policy and the care of children: an analysis using qualitative and quantitative data. Soc. Forces 79 (3), 913–943.

Wang, W., 2016. Genomics and Traditional Chinese Medicine. In: Dhavendra, Kumar., Ruth, Chadwick. (Eds.), Genetics and Society: Ethical, Legal, Cultural and Socioeconomic Implications. Elsevier.

Wang, P., Chen, Z., 2013. Traditional Chinese medicine ZHENG and omics convergence: a systems approach to post-genomics medicine in a global world. OMICS 17 (9), 451–459.

Wang, W., Yan, Y., 2012. Suboptimal health: a new health dimension for translational medicine. Clin. Transl. Med. 1 (1), 28. doi: 10.1186/2001-1326-1-28.

Wang, F., Zuo, X., 1999. Inside China's cities: institutional barriers and opportunities for urban migrants. Am. Econ. Rev. 89 (2), 276–280.

Wang, E., Bussom, S., Chen, J., Quinn, C., Bedognetti, D., Lam, W., et al., 2011. Interaction of a traditional Chinese medicine (PHY906) and CPT-11 on the inflammatory process in the tumor microenvironment. BMC Med. Genomics 4 (1), 38.

Wang, W., Russell, A., Yan, Y., 2014. Traditional Chinese medicine and new concepts of predictive, preventive and personalized medicine in diagnosis and treatment of suboptimal health. EPMA J. 5 (1), 4.

Wang, Y., Adua, E., Russell, A., Roberts, P., Ge, S., Zeng, Q., et al., 2016a. Glycomics and its application potential in precision medicine. Science. doi: 10.1126/science.354.6319.1601-b.

Wang, Y., Klarić, L., Yu, X., Thaqi, K., Dong, J., Novokmet, M., et al., 2016b. The association between glycosylation of immunoglobulin G and hypertension: a multiple ethnic cross-sectional study. Medicine 95 (17).

Wang, Y., Ge, S., Yan, Y., Wang, A., Zhao, Z., Yu, X., et al., 2016c. China suboptimal health cohort study: rationale, design and baseline characteristics. J. Transl. Med. 14 (1), 291.

Wang, Y., Liu, X.X., Qiu, J., Wang, H., Di, Liu., et al., 2017. Association between Ideal Cardiovascular Health Metrics and Suboptimal Health Status in Chinese Population. Scientific Reports 7, 14975. doi: 10.1038/s41598-017-15101-5.

Wasser, S.P., 2002. Medicinal mushrooms as a source of antitumor and immunomodulating polysaccharides. Appl. Microbiol. Biotechnol. 60 (3), 258–274.

World Population Prospects: The 2004 Revision Population Database. New York: United Nations Population Division; 2006.

Yan, S.S., Li, Y., Wang, Y., Shen, S.S., Gu, Y., Wang, H.B., et al., 2008. 17-Acetoxyjolkinolide B irreversibly inhibits IκB kinase and induces apoptosis of tumor cells. Mol. Cancer Ther. 7 (6), 1523–1532.

Yan, Y.X., Liu, Y.Q., Li, M., Hu, P.F., Guo, A.M., Yang, X.H., et al., 2009. Development and evaluation of a questionnaire for measuring suboptimal health status in urban Chinese. J. Epidemiol. 19 (6), 333–341.

Yan, Y.X., Dong, J., Liu, Y.Q., Yang, X.H., Li, M., Shia, G., et al., 2012. Association of suboptimal health status and cardiovascular risk factors in urban Chinese workers. J. Urban Health 89 (2), 329–338.

Yan, N., Chen, N., Lu, J., Wang, Y., Wang, W., 2013. Electroacupuncture at acupoints could predict the outcome of anterior nucleus thalamus high-frequency electrical stimulation in medically refractory epilepsy. Med. Hypotheses 81 (3), 426–428.

Yan, Y.X., Dong, J., Liu, Y.Q., Zhang, J., Song, M.S., He, Y., et al., 2015. Association of suboptimal health status with psychosocial stress, plasma cortisol and mRNA expression of glucocorticoid receptor α/β in lymphocyte. Stress 18 (1), 29–34.

Yang, Y.M., Li, Y., Liu, C.R., 2005. The situations and measures of infant birth qualities from 2000 to 2004. Chin. Primary Health Care 6, 10–11.

Year Book, China, 1991. Beijing: National Bureau of Statistics; 1991. pp. 79–81

Year Book, China, 2008. Beijing: National Bureau of Statistics; 2008. pp. 79–80.

Year Book, China, 2008. Beijing: National Bureau of Statistics; 2008. p. 315

Yu, F., Takahashi, T., Moriya, J., Kawaura, K., Yamakawa, J., Kusaka, K., et al., 2006. Traditional Chinese medicine and Kampo: a review from the distant past for the future. J. Intl. Med. Res. 34 (3), 231–239.

Yue, P.Y., Mak, N.K., Cheng, Y.K., Leung, K.W., Ng, T.B., Fan, D.T., et al., 2007. Pharmacogenomics and the Yin/Yang actions of ginseng: anti-tumor, angiomodulating and steroid-like activities of ginsenosides. Chin. Med. 2 (1), 6.

Yun, H., Song, M.S., Wang, W., 2012a. Traditional Chinese medicine in the area of genomics. In: Dhavendra, K. (Ed.), Genomic and Health in the Developing World. Oxford University Press, pp. 842–845.

Yun, H., Hou, L., Song, M., Wang, Y., Zakus, D., Wu, L., et al., 2012b. Genomics and traditional Chinese medicine: a new driver for novel molecular-targeted personalized medicine? Curr. Pharm. Pers. Med. 10 (1), 16–21.

Zhang, Y.Z., Zhong, N., 2006. Current genetic counseling in China. J. Peking Univ. 38 (1), 33–34.

Zhang, A., Sun, H., Yan, G., Cheng, W., Wang, X., 2013. Systems biology approach opens door to essence of acupuncture. Complement. Ther. Med. 21 (3), 253–259.

Zhao, J., 1999. China Geography. Higher Education Press, Beijing.

Zhao, Y., Cui, S., Yang, J., Wang, W., Guo, A., Liu, Y., Liang, W., 2011. Basic public health services delivered in an urban community: a qualitative study. Public Health 125 (1), 37–45.

Zhao, J., Wang, X., Xu, J., Li, N., Shang, X., He, Z., et al., 2012. Association of inflammatory response gene polymorphism with atherothrombotic stroke in northern Han Chinese. Acta Biochim. Biophys. Sin (Shanghai) 44 (12), 1023–1030.

Zhao, F., Chen, Y., Ge, S., Yu, X., Shao, S., Black, M., et al., 2014. A quantitative analysis of the mass media coverage of genomics medicine in China: a call for science journalism in the developing world. Omics 18 (4), 222–230.

Zheng, S., Song, M., Wu, L., Yang, S., Lu, X., Du, J., et al., 2010a. Public health genomics: China's aspect. Public Health Genomics 13 (5), 269–275.

Zheng, S., Du, J., Lu, X., Zhang, Y., Hu, L., Wang, W., 2010b. Quality of randomized controlled trials in acupuncture treatment of hepatitis B virus infection—a systematic review. Acupunct. Electro-Ther. Res. 35 (3–4), 119–131.

# CHAPTER 4

# Leveraging International Collaborations to Advance Genomic Medicine in Colombia

**Alicia Maria Cock-Rada\*,\*\*, Carlos Andres Ossa Gomez†**
*\*Instituto de Cancerología SA, Medellin, Colombia; \*\*Universidad de Antioquia, Medellin, Colombia; †Clinica Las Américas, Medellin, Colombia*

## INTRODUCTION

The National Human Genome Research Institute defines genomic medicine as "an emerging medical discipline that involves using genomic information about an individual as part of their clinical care (e.g., for diagnostic or therapeutic decision-making) and the health outcomes and policy implications of that clinical use" (NHGRI, 2016). Genomics is now becoming essential in different fields of medicine, such as oncology, neurology, pediatric disorders, and pharmacology. Since the complete sequencing of the human genome in 2003, huge advances have been made in understanding the genetic etiology of diseases and developing more personalized care (Feero et al., 2010; Guttmacher and Collins, 2002). However, there is still a huge gap between high-income and low- to middle-income countries in the use and impact of genomics in regular clinical care. In 2010 the World Health Organization (WHO), published the "Report of a WHO consultation on community genetics in low- and middle-income countries (LMIC)" (WHO, 2010). According to the WHO, the goal of community genetics in LMIC is "to prevent congenital disorders and genetic diseases at population level and, at the same time, to provide genetics services (diagnosis and counseling) in the community for individuals and families." The group of experts consulted by the WHO agreed that availability of community genetic services in LMIC is less than adequate, mainly because of "paucity of resources; genetic conditions not being considered priorities; misconceptions that the control of common congenital disorders is too expensive and linked with sophisticated technology; low genetics literacy; cultural, legal and religious limitations such as the fear of stigmatization within the community and the legal or religious

Genomic Medicine in Emerging Economies.http://dx.doi.org/10.1016/B978-0-12-811531-2.00004-1

restrictions; an insufficient number of trained health professionals; and inadequate data on the true magnitude and economic burden of congenital disorders" (WHO, 2010).

In general, all these limitations are present in most countries in Latin America. In Colombia there was a delay in introducing genomic medicine in standard clinical care, compared with countries such as Argentina, Mexico, and Brazil. One of the main reasons for this is limited health care resources, but also policy makers, health care providers, and health insurances lack a general knowledge of the importance of genetics in medical care. In general, there is poor health planning, insufficient training in genetics for health care professionals, a lack of adequate laboratories and infrastructure, and the molecular diagnostic tests available in the country are very expensive. In Colombia one of the most advanced medical fields in genomic medicine is oncology, in which genetic and genomic tests are starting to be more widely used for cancer patients. To overcome the obstacles mentioned above, a strategy involving different health care actors and international collaborators was developed at the Instituto de Cancerología SA (IDC) in Medellin, Colombia.

## COLOMBIAN POPULATION

Colombia is a country located in the northwestern part of South America. The Spanish explored the Colombian territory for the first time in 1499. In the first half of the 16th century, the Spanish initiated a period of colonization, creating the New Kingdom of Granada, with the capital, Santa Fe de Bogotá. In 1717 the Viceroyalty of New Granada was created and gained its independence from Spain in 1819. After many transitions in its government and territories, it finally became the Republic of Colombia in 1886 (BHEC, 2015).

Colombia has a total land area of 1,138,910 square kilometers, which makes it slightly less than twice the size of Texas. Its gross domestic product (GDP) was 292.08 billion USD in 2015, according to the World Bank (World Bank, 2016). Even though Colombia has an upper-middle income, as defined by the World Bank, it has very high socioeconomic inequalities, similar to other countries in Latin America, reflected by a high Gini index (a measure of income inequality) (Davies et al., 2007; de Andrade et al., 2015). In 2004, the average income Gini index reached 52.5 in Latin America, which was 8 points higher than in Asia, 18 points higher than in Eastern Europe and Central Asia, and 20 points higher than in high-income countries (Gasparini and Lustig, 2011). After 2005, Gini indexes in Latin America started improving (de Andrade et al., 2015). Close to half of the Latin American population lives in poverty and 27.6% earns less than $2 per day, according to the World Bank (World Bank, 2014).

The high degree of socioeconomic disparity leads to wide differences in both health indexes and access to health care services among different socioeconomic groups.

Colombia has an ethnically diverse population, composed of three major groups: the descendants of its indigenous people (Amerindians or Native Americans), European immigrants (mostly Spanish), and Africans originally brought as slaves (Aristizabal, 2000) (Fig. 4.1). Spanish immigrants began colonization of the country in 1510, in San Sebastián de Urabá. In 1526, the Spanish founded Santa Marta on the Caribbean Coast, and in 1533 they founded Cartagena de Indias, which became the main port for trade and for the import of African slaves. Spanish colonists settled mainly in the Andean highlands and the Caribbean coast (Fig. 4.1) (BHEC, 2015) With time, the colonizers established populations in areas higher than 1000 m of altitude to escape mosquito-transmitted diseases, such as malaria or yellow fever. The Spanish conquest was devastating for the indigenous people. It is believed that 150 years after the first Spanish settled, nearly 90% of all Native Americans in the country had died, mainly as a result of diseases brought by Europeans, labor exploitation, armed conflicts with the European settlers, and separation from their families (BHEC, 2015). Since European men came with very few women, they had children with native women, giving rise to mixed or "mestizo" populations that replaced the original natives (BHEC, 2015) In spite of this, Colombia has still over 80 indigenous tribes that speak 66 different languages, making it the second country, after Brazil, with the most indigenous tribes (Aristizabal, 2000). Each tribe is unique in its customs, beliefs, and artistic expressions.

From the early 16th century (starting from 1520) until the 18th century, African slaves were brought by the Spanish and the English from Congo, Angola, Gambia, Senegal, Nigeria, Ivory Coast, Guinea, Sierra Leone, and Mali to increase labor force (Navarrete, 2005). The slaves settled mainly along the Pacific coast, with a minority settling in the Caribbean coast and islands (Fig. 4.1) (Aristizabal, 2000). During this period until the 19th century, there was very little immigration of other populations. Later waves of immigration brought people from the Middle East, Romani populations, Germans, Italians, and Jews, and although these represent only a small minority of the population, they had a deep impact on economic, social, and cultural development in specific regions of the country (Tovar Pinzón, 2001).

The largest wave of immigrants from the Middle East (Syria, Lebanon, Jordan, and Palestine) occurred form 1880 until the 1920s, settling initially on the Caribbean coast (Cartagena and Barranquilla) (Ángel Arango, 1992). Arabs continued to immigrate, and in 1940 there was a large wave of Arab migration to Maicao (border with Venezuela). Jews came initially in the 16th century as "New Christians" and in the 18th century Spanish and Portuguese Jews came

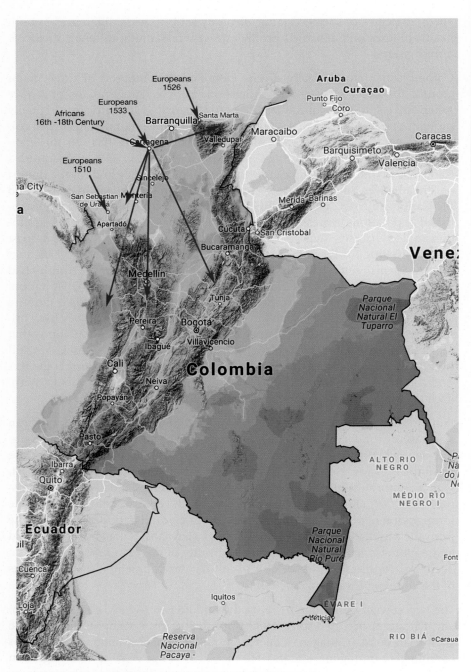

**FIGURE 4.1** Migration history and admixture in Colombia.

from Jamaica and Curacao (Klich and Lesser, 1997). During the 20th century, Sephardic Jews came from Greece, Turkey, North Africa, and Syria, and shortly after they arrived from Eastern Europe. A wave of Ashkenazi Jews came with the rise of Nazism in 1933 and continued to immigrate until the 1950s and 1960s. The Jews ended up being more concentrated in Bogota and Barranquilla in the 21st century. During the 19th and 20th centuries many Germans, some via Venezuela, settled inland and on the Caribbean coast as farmers or professional workers. During and after World War II, numerous Germans came to the country, although many left in the 1980s (SICREMI, 2014).

The genetic composition of the Colombian population reflects the migratory history of Colombia (Fig. 4.1). Admixture mapping studies have shown great genetic heterogeneity among different regions of the country (Adhikari et al., 2016; Ruiz-Linares et al., 2014). Ruiz-Linares et al. (2014) estimated the individual African/European/Native American admixture proportions of Colombians in 1659 by using 30 highly informative single nucleotide polymorphisms (SNP) and found the highest African ancestry in the coastal regions (mainly on the Pacific coast), a higher European ancestry in certain central areas of the country, and the highest Amerindian in the eastern and southwestern parts of the country, including Amazonia (Fig. 4.1). Interestingly, there are populations in the country considered genetic isolates, with a higher incidence of genetic disorders due to geographical isolation, a high degree of consanguinity, inbreeding, or a known founder effect (Arcos-Burgos and Muenke, 2002). The population of Antioquia Province, for example, is composed of 70% European ancestry and is descendant mainly of Spaniards, Sephardic Jews, and Basques, with a low admixture with Amerindian and African populations (Arcos-Burgos and Muenke, 2002; Rojas et al., 2010). A study of this population with genetic markers in the Y chromosome and mitochondrial DNA showed that the origin of men (Y chromosome) is 94% of European ancestry compared with 90% of maternal ancestry (mitochondrial DNA) being

◀ The Spanish colonizers arrived in the early 16th century and founded San Sebastián de Urabá (which later disappeared), Santa Marta, and Bogota. They migrated (*red arrows*) inland and finally settled mainly in the Andean highlands and on the Caribbean coast, annihilating a considerable portion (90%) of the native population (*light orange shade* represents a higher proportion of European ancestry in the current population, >50% European ancestry, and *darker orange areas* represent >70% European ancestry). African slaves were brought to the country from the early 16th century until the 18th century and settled (*green arrows*) mainly along the Pacific coast with a minority settling on the Caribbean coast (*green shade* represents 40%–50% African ancestry, and *darker green areas* represent >50% African ancestry). The southeastern part of the country has mostly Amerindian ancestry (*purple shade* represents >40% Amerindian ancestry and *darker purple areas* >50%). *Modified from Google Maps: https://www.google.com.co/maps/place/Colombia. Data taken from Ruiz-Linares, A., Adhikari, K., Acuña-Alonzo, V., Quinto-Sanchez, M., Jaramillo, C., Arias, W., et al. 2014. Admixture in Latin America: geographic structure, phenotypic diversity and self-perception of ancestry based on 7,342 individuals. PLoS Genet. 10(9), e1004572.*

Native American (Carvajal-Carmona et al., 2000). This revealed the uneven mating pattern in this region in early times, involving mainly immigrant men with local native women, and confirmed what was historically known. Certain genetic conditions, such as metabolic disorders, bipolar disease, chorea, schizophrenia, Tourette syndrome, and cleft palate are common in Antioquia (Arcos-Burgos and Muenke, 2002; Camargo et al., 2012; Kremeyer et al., 2010; Pérez-Póveda et al., 2005; Scharf et al., 2013). A founder mutation in the PSEN1 gene, known to cause a severe familial form of early-onset Alzheimer's disease was also discovered in a specific region of Antioquia (Acosta-Baena et al., 2011).

The genetic background of a population is very important when studying the incidence and mortality of certain diseases, such as breast cancer. In the United States the incidence of breast cancer was shown to be lower in African Americans than in white women, and mortality higher in African Americans (DeSantis et al., 2014). Although differences in exposure to risk factors contribute to the difference in incidence rates, and social and economic disparities can explain the higher rates of mortality in African American women, ethnicity appears to play an important role. In a study conducted by Fejerman et al. (2008) in Latinas living in northern California, they found that a higher European ancestry was associated with increased breast cancer risk, with an odds ratio (OR) of 1.79 for a 25% increase in European ancestry. Another study by Fejerman et al. (2013) showed that a higher indigenous American ancestry was associated with an increased risk of breast cancer-specific mortality (hazard ratio (HR): 1.57 per 25% increase in indigenous American ancestry). A case control study conducted in women residing in Mexico, found that for every 25% increase in European ancestry, there was a 20% increase in risk for breast cancer (Fejerman et al., 2010).

Regarding the study of hereditary cancers, the genetic background of a population should also be taken into account, and sometimes cost-effective screening tests can be developed when founder mutations are responsible for a high proportion of cases in a given population, as is the case with Ashkenazi Jews, where 1 in 40 people are carriers of one of three founder mutations in BRCA1/2 (Ferla et al., 2007).

In Colombia three founder mutations in the genes BRCA1 and BRCA2 (which cause hereditary breast and ovarian cancer syndrome) were discovered in Bogota, and were also found in subsequent studies in breast and ovarian cancer patients in Bogota (Rodríguez et al., 2012; Torres et al., 2007, 2009). A commercial panel test, the "Colombian profile," was developed to study these mutations, and until recently this was the only BRCA1/2 test covered by the health care system in Colombia, leading to incomplete Breast cancer gene testing and inaccurate clinical management of patients and their families. A study in Medellin (the capital of Antioquia Province) in 2011 found a very low frequency of these mutations in unselected breast cancer patients (2%) (Hernández et al., 2014). Recently, our group in the IDC studied breast and ovarian cancer patients with genetic testing criteria, and found a wide spectrum of germline mutations in *BRCA1/2* not previously reported in the country in patients born in different regions, and we also found an unknown recurrent mutation in Antioquia (Cock-Rada et al., 2017). Therefore we demonstrated the need for performing a comprehensive analysis of both genes as well as other cancer-predisposition genes in patients in our population, and the necessity of carrying out more extensive studies in different regions of the country to determine the prevalence and spectrum of mutations in these genes, in order to develop more cost-effective strategies for hereditary cancer testing that could be adopted by the health care system.

With the introduction of genomic medicine in clinical practice, an understanding of the genetic background of the each population is necessary in order to better guide diagnosis and personalized management of diseases with a genetic component.

## HEALTH CARE SYSTEM IN COLOMBIA

Colombia had an estimated population of 49 million people in 2017, making it the third most populous country in Latin America after Brazil and Mexico (DANE, n.d.). The total fertility rate in 2015 was 1.9 births per woman (World Bank, n.d.). Since the 1960's, Colombia experienced a marked decrease in mortality, especially in children, but also in fertility, changing dramatically the demographic composition of the country (Profamilia, 2015). In 2015 in Colombia, 26.8% of the population was younger than 15 years, 65.7% was between 15 and 64 years, and 7.4% was 65 years old or older (Profamilia, 2015). The life expectancy at birth rose from 70.1 years in 2000 to 74.2 years in 2015 for both sexes, being 77.8 years for females and 70.7 years for males (World Bank, 2016).

The health care system was reformed in the 1990s, leading to a considerable improvement in coverage, but increasing health disparities in different aspects. The current social security system was created on December 23, 1993, with the Law 100 (Ley 100) (Congreso, 1993). This system involves three parties: (1) The state (mainly the Ministry of Health and Social Protection and other regulatory entities), which regulates and controls the health care system; (2) the insurance companies or health promoting entities (EPS), which affiliate the population and manage resources; and (3) the health providing institutions (IPS), which include hospitals, clinics, laboratories, and health professionals, who directly treat patients. The health care system is divided into two systems or regimes: (1) the *contributive regime*, financed mainly by employers, employees, and independent workers; and (2) the *subsidized regime* or selection system of beneficiaries for social programs (SISBEN), for people with the lowest socioeconomic level, funded by the state mainly through taxes (Fig. 4.2) (Vargas et al., 2010).

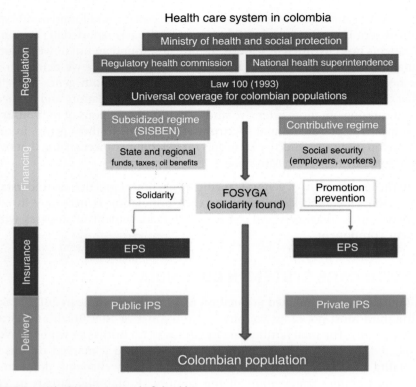

**FIGURE 4.2 Healthcare system in Colombia.**
The health care system in Colombia involves three parties: (1) the state (Ministry of Health and Social Protection and regulatory entities, (2) the insurance companies or health promoting entities (EPS) that affiliate the population, and (3) the health providing institutions (IPS), which include hospitals, clinics, laboratories, and health professionals, who deliver medical care. The health care system is divided into two systems or regimes: the *contributive regime* and the *subsidized regime* or SISBEN. The Solidarity and Guarantee Fund (FOSYGA) finances medical services not included in the Mandatory Health Plan (Plan Obligatorio de Salud, POS). *Modified from http://www.social-protection.org/gimi/gess/RessourcePDF. action?ressource.ressourceId=48019. Data taken form Vargas, I., Vázquez, M.L., Mogollón-Pérez, A.S., Unger, J.-P., 2010. Barriers of access to care in a managed competition model: lessons from Colombia. BMC Health Serv. Res. 10, 1-12.*

The Colombian State created a benefits plan, called the "Mandatory Health Plan (Plan Obligatorio de Salud, POS)," and regulates the health activities and services to which the users of both regimes are entitled, which includes primary, secondary, and tertiary interventions. Individuals who have economic means can choose to pay an additional health insurance called "Prepaid Medicine," in which they choose the health services, and thus they usually have access to the best medical care.

The population coverage of the health care system increased from 37% at the beginning of the 1990s to an estimated 94% in 2015 (Giuffrida et al., 2009).

However, the growth of the system has been uneven, and it is estimated that between 2002 and 2008 the subsidized system grew 2.8 times more than the contributive system (Calderón Agudelo et al., 2011). The Colombian Constitutional Court decided to adjust equally the benefits plan (POS) for both systems, based on the principle of equality. Therefore, health expenditure went from 5.9% in 2000 to 7.2% of GDP in 2014, putting the Colombian health system at risk (World Bank, n.d.).

Since health care resources in the country are limited, genetic disorders have not been a public health priority. The introduction of genetic diagnostic tests in clinical care was slowed down by not being included in the Mandatory Health Plan (POS) until recent years (Resolución 5592 Dic de 2015) (MSPS, 2015). However, since these tests remain very expensive and are mainly performed outside the country, insurance companies usually deny these tests to most patients even if they are obliged to cover their costs. Then, only patients, who demand testing through legal action for "protection of the right to health and life," get the tests paid for by the insurance companies, which eventually are reimbursed by the government. This increases the costs for patients and for the health care system and delays appropriate diagnosis and management. Fortunately, this is currently changing, and genetic tests are being progressively included in Colombian management guidelines for different diseases and are therefore more often financed by the health care system. As mentioned before, one of the fields with a more accelerated advance in genomic medicine in Columbia is oncology, because of the global movement toward a more personalized treatment based on genomic profiles of tumors and the development of targeted cancer therapies. Therefore cancer patients are starting to have more access and benefit from these approaches. However, huge gaps still exist between the subsidized and the contributive systems. Patients in the subsidized system generally receive less and a worse quality of health care, reflected by barriers and delayed access to medical care and late diagnosis and treatment. In the contributive system, there are also great differences among insurance companies (EPS) in access and quality of care. A study performed in the IDC, showed the shocking difference in outcome of breast cancer patients according to the type of insurance held by patients, subsidized or contributive, and even among insurance companies within the contributive regime (EPS) (García et al., 2017). Patients from the contributive regime had a mortality rate of 10% during the follow-up period, and 12% had metastasis or recurrence, compared with a mortality rate of 23% and metastasis or recurrence in 20.6% of patients from the subsidized regime ($P < 0.05$) (García et al., 2017). The time of access to oncologic treatment was twice the time in the subsidized regime (112 days) compared with the contributive regime (52 days), which had an important impact on the prognosis of the patients (García et al., 2017). This illustrates the disparities in health care according to the socioeconomic level and the type of regime.

## CLINICAL GENETICS AND GENETIC COUNSELING IN COLOMBIA

In Colombia there are very few trained clinical or medical geneticists (Javeriana, 2013). There are also wide gaps in access to genetic consultation and genetic counseling according to the region, the type of insurance and insurance company, and the socioeconomic level of the patient. Some cities, such as Bogota and Medellin, have several genetic units mainly in academic institutions, whereas other major cities do not have any genetic services.

One of the main reasons for the shortage of genetic specialists is the lack of genetic training programs in the country for general and specialized medical doctors. There is only one university in Colombia offering a formal clinical genetics residency program, Pontificia Universidad Javeriana, in Bogota (PUJB, n.d.).

Some universities in different cities offer masters and doctoral programs in human genetics or cancer genetics for physicians who also want to pursue a research career. Colombia also has very few PhDs compared with other developing countries, and, specifically, there are very few medical doctors with a PhD degree.

Genetic counseling is an important part of a genetics consultation and is of great support when ordering genetic tests in a clinical setting. In Latin America, clinical geneticists and medical doctors with genetics training usually perform the genetic counseling. The only country that developed genetic counseling as a separate field is Cuba, through a 6-month training program for general physicians (Cruz, 2013; Penchaszadeh, 2004). In Colombia, there are no official training programs for genetic counselors, and unlike in other countries, the insurance companies do not pay the consultations with nonphysician genetic counselors. Therefore only medical doctors are able to officially deliver genetic counseling through the health care system. Sometimes nonphysicians participate in this process in a private practice setting.

There are many research groups across the country with great experience and international publications in different areas of genetics, but most research and training in genetics is in basic sciences (Colciencias, 2016; Corporación Ruta, 2016). Translational and clinical research has only started developing in the last few years, as evidenced by the recent growth of research units in clinics and hospitals conducting clinical research studies financed by pharmaceutical companies or in cooperation with national or international academic institutions.

## GENETICS LAWS AND REGULATIONS IN COLOMBIA

Colombian law mainly addresses three aspects regarding genetics (The Código Penal—Ley 599 del 2000) (Congreso de la República de Colombia, 2000):

1. Genetic manipulation (Article 132): Whoever manipulates human genes with a purpose other than treatment, diagnosis or scientific research in Biology, Genetics and Medicine, oriented at alleviating human suffering or at improving the health of the person of human kind, will be imprisoned 1–5 years.
2. Human being reproducibility (Article 133): Whoever generates identical human beings by cloning or other methods, will be imprisoned 2–6 years.
3. Fecundation and trafficking of human embryos (Article 134): Whoever fertilizes human ovules with a purpose different than human procreation or with prejudice to scientific research, treatment or diagnosis of the person object of the research, will be imprisoned 1–3 years, as well as whoever traffics gametes, zygotes, and human embryos obtained by any means.

Additionally, Colombian law (Ley 721 de 2001) regulates paternity and maternity DNA tests and the authorized laboratories offering these genetic services (Congreso de la República, 2001). However, other genetic tests are not mentioned in Colombian legislation.

A very important aspect, which is not contemplated in Colombian law is genetic discrimination. Ideally, individuals should be protected, as is the case with the Genetic Antidiscrimination Act (GINA) in the United States, against insurance and work discrimination based on genetic conditions (U.S Congress, 2008). Hopefully, this aspect can be addressed in the near future.

Orphan diseases are also addressed in Colombian legislation in the Law 1392/2010 and are defined as chronic, debilitating, and severe medical conditions with prevalence lower than 1 in 5000 people (Congreso de la República de Colombia, 2010). There is a list of all conditions considered orphan diseases that is updated every 2 years (Resolution 0430/2013) and an information system to report and collect information regarding these diseases (Decreto 1954/2012) (Ministerio de Salud y Protección Social, 2013; MSPS, 2012). A special health regulatory commission (Comisión de Regulación en Salud—CRES) was created to evaluate the diagnostic tests that should be included in the benefits plan, POS. The services related to orphan diseases that are not incorporated in the POS, should be financed by a fund called Solidarity and Guarantee Fund (Fondo de Solidaridad y Garantía—FOSYGA) (Law 1392/2010) (Congreso de la República de Colombia, 2010). Newborn or prenatal screening diagnostic tests for genetic disorders are available in the country, but they are not regulated or financed by the health care system. The only mandatory neonatal screening test is for congenital hypothyroidism (Resolution 412/2001, of the Ministry of Health and Social Protection), however, this is not always done, especially in rural areas without access to this test

(MSPS, 2012). Neonatal screening tests for genetic disorders do not necessarily use expensive technologies such as next-generation sequencing (NGS), which have proven to be cost-effective in other countries (Kingsmore et al., 2012). It is estimated that in the near future a population-specific test panel could cost US$10 for 10 Mendelian disorders, which currently affect 2% of children worldwide (Kingsmore et al., 2012). Hopefully, Colombia will understand the importance of regulating genetic tests and making them available to the general population for diagnosing and treating certain diseases, which if recognized early and treated, greatly benefit the patient (e.g., phenylketonuria, galactosemia, and cystic fibrosis).

## MOLECULAR/GENETIC DIAGNOSTIC LABORATORIES

Colombia has a state laboratory, the Instituto Nacional de Cancerología, which has limited funding and resources. Private laboratories perform most molecular/genetic diagnostic tests. According to the Colombian Association of Human Genetics (ACHG), there are at least 40 laboratories offering clinical genetic and cytogenetic testing services, including laboratories in public or academic institutions and over 40 research groups in genetics according to Administrative Department of Science, Technology and Innovation (Colciencias) (Colciencias, 2016). Since there is no regulation for genetic tests in Colombia, besides paternity tests, these are performed in various ways by different laboratories. Many commercial laboratories offer in their portfolios complex genetic tests that are performed abroad, such as multigene panels or whole exome sequencing (WES). They send the patient's samples abroad and receive the results, charging a high amount for acting as a channel between the patient, the health care insurance company, and the laboratory abroad. Other labs send samples to laboratories that offer next-generation sequencing services and charge for the data analysis. However, some of them are research laboratories lacking certifications for diagnosis, such as the Clinical Laboratory Improvement Amendment (CLIA), a certification required in the United States for human diagnostic testing (FDA, 2014).

Some companies in Colombia are starting to invest in sequencing platforms and high-tech infrastructure. However, the high costs of importing the equipment and reagents make the locally performed genetic tests sometimes more expensive than the tests performed abroad. Since there is also a lack of knowledge and of trained professionals (i.e., bio-informaticians, genetic counselors, and geneticists), sometimes the results and especially the analyses and interpretation of results are not very reliable, with a higher risk of false positive or false negative results. A common strategy now is for local labs to partner with laboratories abroad with more experience, to provide local tests with the expertise and support from international labs.

Several laboratories in Colombia are currently offering genetic tests that do not require very expensive equipment and infrastructure to guide personalized medicine in cancer patients, for example, by using RT-PCR or Sanger sequencing to study specific mutations or expression patterns of genes in tumors that confer resistance or sensitivity to certain chemotherapeutic drugs or targeted therapies. Cardona et al. (2011) reported the results of mutation screening of the Epidermal Growth Factor Receptor (*EGFR*) and Kirsten RAt sarcoma virus (*KRAS*) genes in 1939 patients with metastatic non-small-cell lung cancer (NSCLC) patients. They found a positivity rate for *EGFR* mutations of 24.7% (479 pts.) and for *KRAS* mutations of 12.9% (116 pts.). In another study, they found an *EGFR* mutation rate of 25.5% in NSCLC patients, which greatly differed from the mutation rates found in other Latin American countries, such as Peru (51.1%) or Argentina population (14.4%) (Arrieta et al., 2015). Another report on personalized therapy for metastatic colorectal cancer (mCRC ) in Colombia included 202 Colombian patients with mCRC who were treated at the Fundación Santa Fe de Bogotá (FSFB) from March 2009 to March 2013 (Vargas et al., 2014). Overall survival (OS) was analyzed according to *KRAS*, v-Raf murine sarcoma viral oncogene homolog B (*BRAF*) and Phosphatidylinositol-4,5-bisphosphate 3-kinase (*PI3K*) mutational profile. They found that 32.7% of the tumors harbored a *KRAS* mutation, 10.9% harbored a *PIK3CA* mutation, and 6.4% harbored a *BRAF* mutation. OS for the *BRAF* wild-type population was 30 months (95% CI 20.1–41.0), whereas those carrying the *BRAF* mutation lived 13 months (95% CI 9.2–17.0; $P = .09$). They concluded that routine genotyping is necessary to improve patient selection for targeted treatments.

With the rapid evolution in sequencing techniques and the constant decrease in the price of molecular/genetic testing technologies, hopefully these services will become more accessible to everyone.

## ONCOGENETICS UNIT IN THE INSTITUTO DE CANCEROLOGÍA LAS AMÉRICAS

The Instituto de Cancerología SA (IDC), an IPS, is a comprehensive cancer center that serves a population of 4 million people of the Medellin metropolitan area and is one of the main referral institutions from other regions of Antioquia and the rest of country. The IDC has a multidisciplinary team composed of oncologists, surgeons, radiotherapists, radiologists, pathologists, palliative care doctors, nurses, a geneticist, epidemiologists, and a psychologist, who provide integrated cancer care. One of the main focal points of the IDC is breast cancer: 800–1000 patients are treated in the IDC every year. The IDC has been a sister institution of MD Anderson since 2011 and actively participates in the Global Academic Program run by the MD Anderson.

The IDC established the first oncogenetics unit in Medellin, in February 2015, for the study of hereditary, familial, and early-onset tumors. Of the patients referred to the oncogenetics unit 80%–90% have breast and/or ovarian cancer. When a germ line mutation in a highly penetrant gene is found in an individual and a hereditary cancer is diagnosed, this greatly impacts the clinical management of the patient and family. For example, germ line mutations in the *BRCA1* and *BRCA2* genes cause hereditary breast and ovarian cancer syndrome, which confer a cumulative lifetime risk for developing breast cancer of 40%–80% and for ovarian cancer of 11%–40% (Petrucelli et al., 2013). These mutations are transmitted in an autosomic dominant manner; therefore, first-degree relatives have a 50% chance of carrying a mutation. Because of the high risks of developing cancer in mutation carriers, prophylactic surgeries and/or strict surveillance significantly increases the survival of these patients (Daly et al., 2017).

We developed a care model for hereditary cancers that integrates the health care system and research projects (Fig. 4.3). The patient referred to oncogenetics consultation with personal or family history of cancer is first seen by a trained nurse who does a preconsultation with the patient, who answers a cancer risk questionnaire developed in the IDC and does a complete family pedigree. Then, the geneticist evaluates the patient, performs genetic counseling, and if needed orders genetic tests. Between 2015 and 2017, only 30% of the tests ordered through the health care system in the IDC for the study of hereditary cancers (based on the National Comprehensive Cancer Guidelines (NCCN)) was covered by the insurance companies (EPS). The test orders were analyzed by a committee created by each insurance company called the Technique-Scientific Committee (CTC), which decided if they would pay for these tests or not. Most tests were denied, and the patient had to sue the insurance companies through a legal mechanism created by the Colombian Constitution called "Tutela Action," which protects against violation of individuals fundamental rights (Presidencia de la República de Colombia, 1991). Therefore results usually took 3–6 months on average to be obtained. This changed in 2017; CTC's were officially eliminated, and insurance companies are obliged to cover all tests included in the POS (e.g., molecular studies for diseases, gene sequencing, deletion and duplication analysis). Nevertheless, many EPS still do not comply with their obligations, resulting in the same legal actions and delayed medical attention.

## INTERNATIONAL COLLABORATIONS TO ADVANCE MEDICAL GENOMICS IN COLOMBIA

Because of the difficulties discussed, many public and private academic and health care institutions use research projects to address problems not covered or ignored by the health care system. Research funding from the state, mainly

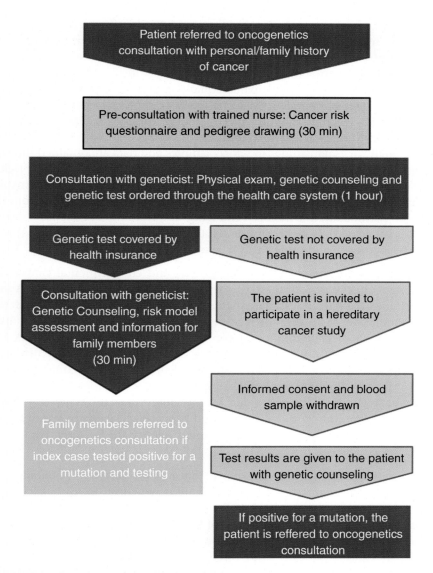

**FIGURE 4.3 Oncogenetics unit model in the Instituto de Cancerología Las Américas.**
The patient is referred to the oncogenetics consultation with personal or family history of cancer. *Red boxes* indicate the steps covered by the health insurance, and *purple* indicates the steps covered by the research projects. The *grey* step is covered by IDC, and *pink* depends on the EPS of the patient.

through Colciencias and the General System of Royalties (from oil companies), was considerably reduced in recent years (Universidad del Rosario, 2017). Therefore an important option is to establish collaborations with institutions abroad and apply for international research funding. The ideal collaboration occurs when local and foreign participants equally gain from a cooperation, which sometimes does not happen. There are many examples of research projects in which foreign researchers or institutions "collect" samples of a less studied population and "disappear". Popular terms for these types of projects are "helicopter research" or "safari study" in which samples and data are obtained through local contacts, often without considering the patient's interests and/or rights. It was even common not to have approval by an Ethics Comittee or an Institutional Review Board (IRB), or not obtain informed, written consent from the patient. Fortunately, this has been changing, and Colombian Laws for Research in Human subjects (e.g., Resolution 8430/1993 and Decreto 2378/2008) now rigorously control and protect this type of research in Colombia.

In order to understand more about genetic and environmental cancer risk in patients in the region, the IDC partnered with the Universidad de Antioquia (UdeA), a public academic higher education institution, to characterize our population. With their own resources or with funding from other institutions, research collaborations have helped develop the oncogenetics and cancer genetics fields in the clinical setting. Several projects are currently being developed by the IDC such as the Study of Familial and Hereditary Cancers in Colombian Population (Funded mostly by City of Hope), the Study of Hereditary Gastric Cancer (funded by the UdeA and University of California, Davis), and the Study of Genetic and Exogenous Factors in Women with Breast Cancer (funded by the International Agency for Research on Cancer (IARC), Colciencias, and UdeA).

The "Study of Familial and Hereditary Cancers in Colombian Population" is an example of an international collaboration beneficial to all participating parties. This project is funded by City of Hope, the IDC, and the University of Antioquia (UdeA). The main goal of the study is to identify genetic alterations in our population that predispose to cancer as well as other factors that modulate cancer risk. This study offers the possibility of initially testing the *BRCA1* and *BRCA2* genes by using a panel of mutations previously reported in different Hispanic populations, the HISPANEL (Abugattas et al., 2015; Alemar et al., 2016; Villarreal-Garza et al., 2015). If no mutations are found, then a complete analysis of both genes is performed as well as of other cancer-predisposing genes. This study is part of an international consortium, the "Clinical Cancer Genetics Research Network (CCGCRN)" directed by Dr. Jeffrey Weitzel of City of Hope in the United States. With the idea of learning more about cancer predisposition and modifying factors in different populations, both germ line and tumor samples will be studied in the future, with the written, informed consent of the patient. This protocol protects the rights and confidentiality of every subject from any third party and was approved by a local independent ethics committee (i.e., Institutional Review Board). The patients are benefiting directly from this study, because test results that change clinical management are disclosed to the patient with genetic counseling. Therefore patients that did not have access to these tests through the health care system and have a clear need for genetic testing according to NCCN guidelines have the opportunity to get tested. One of the main advantages for the collaborating sites is the academic feedback and the possibility to interact and discuss cases through a weekly Web meeting or through case discussion forums. City of Hope also offers a world-renowned training program, the City of Hope Clinical Cancer Genomics Community of Practice (CCGCoP) Intensive Course in Cancer Risk Assessment, which is partly funded by the National Cancer Institute. This course is composed of multidisciplinary participants (e.g., genetic counselors, geneticists, oncologists, surgeons, and biologists) from different parts of the world and offers physicians, geneticists, and genetic counselors form all over Latin America basic training in cancer risk assessment, which is not available in their own countries. City of Hope, along with the University of Chicago, also offers yearly academic events and update conferences on cancer genetics and genomics.

A study of hereditary diffuse gastric cancer was also developed in collaboration with the University of Antioquia and the Genome Center and the Department of Biochemistry and Molecular Medicine in the School of Medicine at the University of California, Davis, directed by Dr. Luis Carvajal-Carmona. We are also participating in the Gastric Cancer Genetics Consortium in Hispanic Populations, which will hopefully help us understand more about this disease in our population, because gastric cancer is a leading cause of cancer-related mortality in the country. We identified individuals with testing criteria for hereditary diffuse gastric cancer and performed E-Cadherin coding gene (CDH1) germ line testing along with DNA repair genes involved in homologous recombination. This study offered the opportunity to include patients, not tested through the health care system, who needed genetic testing. This collaboration also offered the opportunity for a master's student in genetics at the University of Antioquia to be trained in next-generation sequencing technologies at the University of California, Davis, as well as to receive bioinformatics training in human genetic analysis, which is very deficient in the country.

Together, these two studies have opened a window of opportunity for the advancement of research in cancer genetics and offer the possibility to study patients ignored by the health care system. The results of these studies will hopefully have an impact on the health care system and policy makers by demonstrating the impact of genetic testing on management of hereditary cancers, which account for about 10% of all cancers.

## CONCLUSION

Given the geographical location of Colombia, different migrations have made it a country with high admixture and genetically heterogeneous populations, which has to be taken into account when addressing diseases with a genetic component. The health care system is slowly introducing genomic medicine into standard medical care, although disparities in the system, the deficiency in medical genetics expertise, and technological infrastructure, as well as the lack of regulation in genetics services and testing, are still the main problem that prevents adequate development of genomic medicine in the country. Important international collaborations have greatly helped the Instituto de Cancerología SA, a comprehensive cancer center in Medellin, to improve systems of diagnosis and treatment for cancer patients and to encourage clinicians to develop medical, technical, and scientific expertise in accordance with international standards. We hope that other institutions in the country and in Latin America learn from our experience at the IDC so that access to genetic medicine can be made available throughout the continent. Hopefully, personalized medicine, comprising genomic tools and the right expertise to treat devastating diseases such as cancer, will become accessible to most of the population in the near future.

## Acknowledgments

We thank the Instituto de Cancerologia SA, Universidad de Antioquia, and City of Hope for their support.

## References

Abugattas, J., Llacuachaqui, M., Allende, Y.S., Velásquez, A.A., Velarde, R., Cotrina, J., et al., 2015. Prevalence of BRCA1 and BRCA2 mutations in unselected breast cancer patients from Peru. Clin. Genet. 88 (4), 371–375.

Acosta-Baena, N., Sepulveda-Falla, D., Lopera-Gómez, C.M., Jaramillo-Elorza, M.C., Moreno, S., Aguirre-Acevedo, D.C., et al., 2011. Pre-dementia clinical stages in presenilin 1 E280A familial early-onset Alzheimer's disease: a retrospective cohort study. Lancet Neurol. 10 (3), 213–220.

Adhikari, K., Mendoza-Revilla, J., Chacón-Duque, J.C., Fuentes-Guajardo, M., Ruiz-Linares, A., 2016. Admixture in Latin America. Curr. Opin. Genet. Dev. 41, 106–114.

Alemar, B., Herzog, J., Brinckmann Oliveira Netto, C., Artigalás, O., Schwartz, I.V., Matzenbacher Bittar, C., et al., 2016. Prevalence of Hispanic BRCA1 and BRCA2 mutations among hereditary breast and ovarian cancer patients from Brazil reveals differences among Latin American populations. Cancer Genet. 209 (9), 417–422.

Ángel Arango, Biblioteca Luis, 1992. En la tierra de las oportunidades: los sirio-libaneses en Colombia La. Boletín Cultural y Bibliográfic 29 (29), 1–3.

Arcos-Burgos, M., Muenke, M., 2002. Genetics of population isolates. Clin. Genet. 61 (4), 233–247.

Aristizabal, S., 2000. La diversidad étnicay cultural de Colombia: Un desafío para la educación. Pedagogía y Saberes (15), 59–66.

Arrieta, O., Cardona, A.F., Martín, C., Más-López, L., Corrales-Rodríguez, L., Bramuglia, G., et al., 2015. Updated frequency of EGFR and KRAS mutations in nonsmall-cell lung cancer in Latin America: the Latin-American Consortium for the investigation of lung cancer (CLICaP). J. Thorac. Oncol. 10 (5), 838–843.

2015. Breve Historia Económica de Colombia, first ed. Fundación Universidad de Bogotá Jorge Tadeo Lozano, Bogotá, Colombia.

Calderón Agudelo, C., Botero Cardona, J., Bolaños Ortega, J., Martínez Robledo, R., 2011. Sistema de salud en Colombia: 20 años de logros y problemas. Ciência & Saúde Coletiva 16 (6), 2817–2828.

Camargo, M., Rivera, D., Moreno, L., Lidral, A.C., Harper, U., Jones, M., et al., 2012. GWAS reveals new recessive loci associated with non-syndromic facial clefting. Eur. J. Med. Genet. 55 (10), 510–514.

Cardona, A.F., Ramos, P.L., Duarte, R., Carranza, H., Castro, C.J., Lema, M., et al., 2011. Screening for mutations in Colombian metastatic non-small-cell lung cancer (NSCLC) patients (ONCOLGroup). J. Clin. Oncol., 29.

Carvajal-Carmona, L.G., Soto, I.D., Pineda, N., Ortíz-Barrientos, D., Duque, C., Ospina-Duque, J., et al., 2000. Strong Amerind/white sex bias and a possible Sephardic contribution among the founders of a population in northwest Colombia. Am. J. Hum. Genet. 67 (5), 1287–1295.

Cock-Rada, A.M., Ossa, C.A., García, H.I., Gómez, L.R., 2017. A multi-gene panel study in hereditary breast and ovarian cancer in Colombia. Fam. Cancer.

Colciencias, 2016. Publicación de Resultados Finales de la Convocatoria de Medición de Grupos 737 de 2015. pp. 1–345.

Congreso, de la República de Colombia, 1993. Ley 100 de 1993, pp. 1–500.

Congreso de la República de Colombia, 2000. Código Penal Colombiano. pp. 1–318.

Congreso de la República de Colombia, 2010. Ley 1392 del 2010. pp. 1–9.

Congreso de la República, 2001. Ley 721 de 2001. pp. 1–9.

Corporación Ruta N, 2016. Informe No. 1 Área de oportunidad Terapia Génica. pp. 1–79.

Cruz, A.L., 2013. An overview of genetic counseling in Cuba. J. Genet. Couns. 22 (6), 849–853.

Daly, M.B., Pilarski, R., Berry, M., Buys, S.S., Farmer, M., Friedman, S., et al., 2017. NCCN Guidelines insights: genetic/familial high-risk assessment: breast and ovarian, version 2.2017. J. Natl. Compr. Canc. Netw. 15 (1), 9–20.

DANE. La población proyectada en Colombia. Available from: http://www.dane.gov.co/reloj/.

Davies, J.B., Sandström, S., Shorrocks, A., Wolff, E.N., 2007. Estimating the Level and Distribution of Global Household Wealth. UNU-WIDER, 1-56.

de Andrade, L.O., Pellegrini Filho, A., Solar, O., Rígoli, F., de Salazar, L.M., Serrate, P.C., et al., 2015. Social determinants of health, universal health coverage, and sustainable development: case studies from Latin American countries. Lancet 385 (9975), 1343–1351.

DeSantis, C., Ma, J., Bryan, L., Jemal, A., 2014. Breast cancer statistics 2013. CA Cancer J. Clin. 64 (1), 52–62.

FDA, 2014. Clinical Laboratory Improvement Amendments (CLIA). Available from: https://www.fda.gov/MedicalDevices/DeviceRegulationandGuidance/IVDRegulatoryAssistance/ucm124105.htm%C2%A02014.

Feero, W.G., Guttmacher, A.E., Collins, F.S., 2010. Genomic medicine—an updated primer. N. Engl. J. Med. 362, 2001–2011.

Fejerman, L., John, E.M., Huntsman, S., Beckman, K., Choudhry, S., Perez-Stable, E., et al., 2008. Genetic ancestry and risk of breast cancer among US Latinas. Cancer Res. 68 (23), 9723–9728.

Fejerman, L., Romieu, I., John, E.M., Lazcano-Ponce, E., Huntsman, S., Beckman, K.B., et al., 2010. European ancestry is positively associated with breast cancer risk in Mexican women. Cancer Epidemiol. Biomarkers Prev. 19 (4), 1074–1082.

Fejerman, L., Hu, D., Huntsman, S., John, E.M., Stern, M.C., Haiman, C.A., et al., 2013. Genetic and risk of mortality among U.S. Latinas with breast cancer. Cancer Res. 73 (24), 7243–7253.

Ferla, R., Calò, V., Cascio, S., Rinaldi, G., Badalamenti, G., Carreca, I., et al., 2007. Founder mutations in BRCA1 and BRCA2 genes. Ann. Oncol. 18 (Suppl. 6), vi93–vi98.

García, H.I., Egurrola, J.A., Gómez, L.R., Herazo, F., Sánchez, V., Ossa, C.A., 2017. Efecto del aseguramiento en salud sobre la supervivencia global y libre de enfermedad de pacientes operadas por cáncer de mama en un centro oncológico de Medellín. Estudio de cohorte histórica. Revista Colombiana de Cancerología 21 (1), 60.

Gasparini, L., Lustig, N., 2011. The Rise and Fall of Income Inequality in Latin America. Society for the Study of Economic Inequality, 1-28.

Giuffrida, A., Flórez, C.E., Giedion, Ú., Cueto, E., López, J.G., Glassman, A., et al., 2009. From Few to Many: Ten Years of Health Insurance Expansion in Colombia. Inter-American Development Bank, Washington, DC, Salud al alcance de todos: Una década de expansión del seguro médico en Colombia.

Guttmacher, A.E., Collins, F.S., 2002. Genomic medicine—A primer. N. Engl. J. Med. 347 (19), 1512–1520.

Hernández, J.E., Llacuachaqui, M., Palacio, G.V., Figueroa, J.D., Madrid, J., Lema, M., et al., 2014. Prevalence of BRCA1 and BRCA2 mutations in unselected breast cancer patients from Medellín, Colombia. Hered. Cancer Clin. Pract. 12 (1), 11.

Javeriana, P.U., 2013. Cendex. Estudio de Disponibilidad y Distribución de la Oferta de Médicos Especialistas, en Servicios de Alta y Mediana Complejidad en Colombia. pp. 1–135.

Kingsmore, S.F., Lantos, J.D., Dinwiddie D.L., 2012. Next-generation community genetics for low- and middle-income countries. Genome Med. 4 (3), 25.

Klich, I., Lesser, J., 1997. Arab and Jewish Immigrants in Latin America: Images and Realities. Psychology Press, pp. 76–78.

Kremeyer, B., García, J., Müller, H., Burley, M.W., Herzberg, I., Parra, M.V., et al., 2010. Genome-wide linkage scan of bipolar disorder in a Colombian population isolate replicates Loci on chromosomes 7p21-22, 1p31, 16p12 and 21q21-22 and identifies a novel locus on chromosome 12q. Hum. Hered. 70 (4), 255–268.

Ministerio de Salud y Protección Social, 2013. Resolución Número 0430 del 2013. pp. 1–38.

Ministerio de Salud y Protección Social, 2012. Decreto 1954 de 2012. pp. 1–4.

Ministerio de Salud y Protección Social, 2015. Resolución Número 5592 del 2015. pp. 1–220.

Navarrete, M.C., 2005. Génesis y desarrollo de la esclavitud en Colombia siglos XVI y XVII, Primera ed. Universidad del Valle, Cali, Coluumbia.

NHGRI, 2016. National Human Genome Research Institute. https://www.genome.gov/27552451/what-is-genomic-medicine/.

Penchaszadeh, V.B., 2004. Genetic services in Latin America. Commun. Genet. 7 (2–3), 65–69.

Pérez-Póveda, J.C., Palacio, L.G., Arcos-Burgos, M., 2005. Description of an endogamous, multi-generational and extensive family with benign hereditary chorea from the Paisa community. Rev. Neurol. 41 (2), 95–98.

Petrucelli, N., Daly, M.B., Feldman, G.L., 2013. BRCA1 and BRCA2 hereditary breast and ovarian cancer. GeneReviews, Available from: https://www.ncbi.nlm.nih.gov/books/NBK1247/.

Presidencia de la República de Colombia, 1991. Decreto 2591 de 1991. pp. 1–12.

Profamilia, Ministerio de Salud y Protección Social, 2015. Encuesta Nacional de Demografía y Salud Resumen Ejecutivo. 2015-1-96.

Pontificia Universidad Javeriana Bogotá. Especialización en Genética Médica. Available from: http://www.javeriana.edu.co/especializacion-genetica-medica.

Rodríguez, A.O., Llacuachaqui, M., Pardo, G.G., Royer, R., Larson, G., Weitzel, J.N., et al., 2012. BRCA1 and BRCA2 mutations among ovarian cancer patients from Colombia. Gynecol. Oncol. 124 (2), 236–243.

Rojas, W., Parra, M.V., Campo, O., Caro, M.A., Lopera, J.G., Arias, W., et al., 2010. Genetic make up and structure of Colombian populations by means of uniparental and biparental DNA markers. Am. J. Phys. Anthropol. 143 (1), 13–20.

Ruiz-Linares, A., Adhikari, K., Acuña-Alonzo, V., Quinto-Sánchez, M., Jaramillo, C., Arias, W., et al., 2014. Admixture in Latin America: geographic structure, phenotypic diversity and self-perception of ancestry based on 7,342 individuals. PLoS Genet. 10 (9), e1004572.

Scharf, J.M., Yu, D., Mathews, C.A., Neale, B.M., Stewart, S.E., Fagerness, J.A., et al., 2013. Genome-wide association study of Tourette's syndrome. Mol. Psychiatry 18 (6), 721–728.

SICREMI, OEA, 2014. Colombia—Síntesis histórica de la migración internacional en Colombia.

The World Bank, n.d. World Bank open data. Available from: https://data.worldbank.org/indicator/SP.DYN.TFRT.IN?locations=CO&view=chart.

Torres, D., Rashid, M.U., Gil, F., Umana, A., Ramelli, G., Robledo, J.F., et al., 2007. High proportion of BRCA1/2 founder mutations in Hispanic breast/ovarian cancer families from Colombia. Breast Cancer Res. Treat. 103 (2), 225–232.

Torres D. UA, Robledo J. F., Caicedo, J.J., Quintero E., Orozco A., Torregrosa L., Tawil M., Hamann U., Briceño, I., 2009. Estudio de factores genéticos para cáncer de mama en Colombia. Bogotá, Colombia.

Tovar Pinzón, H., 2001. Emigración y éxodo en la historia de Colombia. Amérique Latine Histoire et Mémoire. Les Cahiers ALHIM, pp. 1–11.

U.S Congress, Genetic Information Nondiscrimination Act of 2008. Available from: https://www.congress.gov/bill/110th-congress/house-bill/00493.

Universidad del Rosario, 2017. Foro sobre financiación de la ciência tecnología e innovación en Colombia. Available from: http://www.urosario.edu.co/Periodico-Nova-Et-Vetera/Actualidad-Rosarista/En-Colombia-los-recursos-para-la-ciencia-vienen-ca/.

Vargas, I., Vázquez, M.L., Mogollón-Pérez, A.S., Unger, J-P., 2010. Barriers of access to care in a managed competition model: lessons from Colombia. BMC Health Serv. Res. 10, 1–12.

Vargas, C.A., Carranza, H., Otero, J.M., Becerra, H.A., Acevedo, A., Rodríguez, J., et al., 2014. Personalized therapy for metastatic colorectal cancer in Colombia (ONCOLGroup). J. Clin. Oncol., 32.

Villarreal-Garza, C., Alvarez-Gómez, R.M., Pérez-Plasencia, C., Herrera, L.A., Herzog, J., Castillo, D., et al., 2015. Significant clinical impact of recurrent BRCA1 and BRCA2 mutations in Mexico. Cancer 121 (3), 372–378.

World Health Organization (WHO), 2010. Community Genetics Services: Report of a WHO Consultation on Community Genetics in Low- and Middle-Income Countries. World Health Organization.

World Bank Group, 2014. Annual Report 2014. World Bank Group, pp 1–67.

The World Bank, 2016. Colombia. Available from: http://data.worldbank.org/country/colombia.

# CHAPTER 5

# Screening for Hereditary Cancer in Latin America

**Karin Alvarez\*, Carolina Alvarez\*\*, Mev Domínguez†, Pilar Carvallo\*\***

*\*Laboratory of Oncology and Molecular Genetics, Coloproctology Unit, Clinica las Condes, Santiago, Chile; \*\*Department of Cell and Molecular Biology, Faculty of Biological Sciences, Pontificia Universidad Católica de Chile, Santiago, Chile; †Department of Tumor Biology, Institute for Cancer Research, Oslo University Hospital, Oslo, Norway*

## INTRODUCTION

Human Molecular Genetics studies on diverse hereditary diseases started in Latin America in the 1990s with few countries having a well-developed Molecular Biology at the time. Geneticists from Mexico, Colombia, Brazil, Argentina, and Chile gathered at the first association, the Latin American Program of the Human Genome (PLAGH). Later in 2000, a second initiative replaced PLAGH and created RELAGH, which is the Latin American Network in Human Genetics, in which most South American countries participate, in addition to Mexico, Cuba, and Costa Rica. RELAGH organizes a meeting every two years with the participation of several countries' representatives. The long distances among many Latin American countries and the scarcity of funds for science research in most places have made it difficult to congregate a high number of scientists at our meetings, as well as to establish collaborations. Regardless all the difficulties, nowadays, several countries have respectable and productive research programs in Molecular Genetics focused either on genetic diseases or on hereditary cancer. The first studies based on screening for mutations in candidate genes in hereditary cancer were carried out by electrophoretic analyses such as Single Strand Conformation Polymorphisms (SSCPs), heteroduplex

Genomic Medicine in Emerging Economies. http://dx.doi.org/10.1016/B978-0-12-811531-2.00005-9

or other equivalent, and Protein Truncation Test (PTT), supported by Sanger sequencing in specific fragments. Full sequencing by Sanger has been widely used in mutational screening since about 2008 to 2010, and in the last 3 to 4 years next-generation sequencing has become more commonly used. In this regard, the countries having installed this latter technology are Mexico, Brazil, Chile, Argentina and Uruguay. One of the main problems in using our own NGS facilities is that the higher prices of the reagents, shipping and taxes, in addition to a low number of analyzed samples make NGS more expensive than in developed countries.

In Latin America, research in hereditary cancer has been mainly focused on breast and colon cancer because of the widely available information relating to the genetics of these two types of cancer and the high occurrence of these malignancies as well.

## HEREDITARY BREAST CANCER

### Clinical and Genetic Features

Since the identification of the two genes *BRCA1* and *BRCA2* (Hall et al., 1990; Miki et al., 1994; Wooster et al., 1995) mutations of which confer a high risk of breast cancer many populations have been screened by selecting women having a strong family history of breast and ovarian cancer. In the first few years (1995–2000), the criteria for selecting patients for *BRCA1* and *BRCA2* screening included only first- degree relatives affected by breast or ovarian cancer. Over the years, several studies in which selection criteria of the patients' were expanded to second degree relatives and isolated patients diagnosed under 40 years, among others, revealed that additional breast cancer patients must be screened for mutations in *BRCA1* and *BRCA2*. For this reason, different studies published from Latin American countries, since the identification of *BRCA1* and *BRCA2*, show different selection criteria of breast cancer patients (Table 5.1). In addition to differences in selection criteria a variety of experimental approaches have been utilized, in relation to screening techniques and the analysis of selected exons or total exons of *BRCA1* and *BRCA2* (Table 5.1). In this chapter, we will present studies carried on mainly within Latin America in which a full screening of *BRCA1* and *BRCA2* has been accomplished, and we will discuss on the different screening technologies used, as well as the criteria for breast cancer patients'. Even if penetrance of mutations at 70 years old has been estimated between 52% and 47% for *BRCA1* and *BRCA2*, respectively, for other populations (Chen and Parmigiani, 2007; Milne et al., 2008), no studies have been performed in this regard in Latin American high-risk families. This is a very important issue since genetic and nongenetic features may influence penetrance.

**Table 5.1** *BRCA1* and *BRCA2* Mutation Screening in Latin American Hereditary Breast Cancer Cases

| Study Population | Number of Families | Family history Features | Screening Method | Families With Mutation (%) | | Number of Mutations | | Number of Recurrent Mutations (% of total carriers) | Number of Founder Mutations (% of total carriers) | Rearrangements |
|---|---|---|---|---|---|---|---|---|---|---|
| | | | | BRCA1 | BRCA2 | Total | Novel | | | |
| Argentina Solano et al. (2012) | 40 | 40 (Ashkenazi-Jewish descent) with two or more relatives with breast cancer and/or presence of relatives with ovarian cancer | Sanger Sequencing of 3 founder mutations | 9/40 (22.5%) | 8/40 (20%) | 3 | 0 | 3 (42.5%) | ND | ND |
| | 57 | Two or more relatives with breast cancer. Presence of relatives with ovarian cancer | Sanger Sequencing | 10/57 (17.5%) | 5/57 (8.8%) | 24 | 6 | 3 (17.5%) | ND | ND |
| Brazil Carraro et al. (2013) | 16 | Early-onset, and family history according to NCCN guidelines | Sanger Sequencing | 6/16 (37.5%) | 1/16 (6.5%) | 6 | 2 | 1 (28.6%) | ND | ND |
| Chile Gallardo et al. (2006) | 54 | Two or more relatives with breast cancer. Presence of relatives with ovarian cancer. Presence of relatives with male breast cancer | PTT, SSCP | 4/54 (7.5%) | 7/54 (13%) | 8 | 2 | 3 (54.5%) | ND | ND |
| Chile Jara et al. (2006) | 61 | Two or more relatives with breast cancer. Presence of relatives with ovarian cancer. Presence of relatives with male breast cancer | CSGE | 7/61 (11.5%) | 3/61 (4.9%) | 9 | 3 | 1 (20%) | ND | ND |

*(Continued)*

**Table 5.1** *BRCA1* and *BRCA2* Mutation Screening in Latin American Hereditary Breast Cancer Cases (*Cont.*)

| Study Population | Number of Families | Family history Features | Screening Method | Families With Mutation (%) | | Number of Mutations | | Number of Recurrent Mutations (% of total carriers) | Number of Founder Mutations (% of total carriers) | Rear-rangements |
|---|---|---|---|---|---|---|---|---|---|---|
| | | | | BRCA1 | BRCA2 | Total | Novel | | | |
| Chile Ewald et al. (2016) | 74 | Two or more relatives with breast cancer. Presence of relatives with ovarian cancer. Presence of relatives with male breast cancer | MLPA | ND | ND | ND | ND | ND | ND | 3 (non pathogenic) |
| Colombia Torres et al. (2007) | 53 | Two or more relatives with breast cancer. Presence of relatives with ovarian cancer. Presence of relatives with male breast cancer | SSCP, DHPLC, PTT | 8/53 (15.1%) | 5/53 (9.4%) | 6 | 1 | 3 (76.9%) | 3 (76.9%) | 0 |
| Uruguay Delgado et al. (2011) | 42 | Two or more relatives with breast cancer. Presence of relatives with ovarian cancer. Presence of relatives with male breast cancer | HDA, PTT | 2/42 (4.8%) | 5/42 (12%) | 7 | 3 | 0 | ND | ND |
| Venezuela Lara et al. (2012) | 51 | Two or more relatives with breast cancer. Presence of relatives with ovarian cancer. Presence of relatives with male breast cancer | CSGE | 4/51 (7.8%) | 3/51 (5.9%) | 6 | 2 | 1 (28.6%) | ND | ND |

*Number of families and mutation frequencies are referred only to hereditary cases in each manuscript.*
*CSGE, Conformation sensitive gel electrophoresis; DGGE, denaturing gradient gel electrophoresis; DHPLC, denaturing high performance liquid chromatography; HAD, heteroduplex analysis; HRM, high resolution melting; MLPA, multiple ligation probe amplification; ND, not determined; NGS, next generation sequencing; PTT, protein truncation test; SSCP, single stranded conformational polymorphism.*

## Amerindian or European Ancestry and Breast Cancer Incidence

It has been demonstrated that breast cancer prevalence is variable depending on the ancestry composition of populations. For example, Asian countries have the lowest incidence rates of breast cancer (20–30/100,000 women), and Caucasian populations such as Europe (80–100/100,000 women) and North America (120/100,000 women) have the highest (GLOBOCAN 2012). In relation to Latin American populations the incidence of breast cancer ranges from 24/100,000 to 71/100,000, with Argentina, Uruguay and Brazil having the highest incidences. It is interesting that the Hispanic population living in the United States has a higher breast cancer incidence (91.9/100,000) than those living in their countries of origin, but lower that Non-Hispanic whites (128.1/100,000). According to GLOBOCAN, the SEER Cancer Statistics Review (Howlader et al., 2016) states that US women of European ancestry have a greater risk of breast cancer than US women with indigenous ancestry. These differences may be caused by genetic and/or nongenetic factors to which these populations are exposed. In this context, it has been demonstrated by Fejerman et al. (2008) that European ancestry constitutes a relevant genetic risk factor to breast cancer, among US Hispanics, with an OR = 1.79 for every 25% increase in European ancestry. In addition, the ancestry analysis for Mexican women residing in Mexico showed that for every 25% increase in European ancestry there was a 20% increase in risk (Fejerman et al., 2010). In this context Uruguay, which has the highest incidence of breast cancer (71/100,000) in South America, has an average European ancestry of 76.6% (Bonilla et al., 2015). Bonilla et al. (2015) found a positive association of breast cancer in Uruguayan women with European ancestry (mitochondrial haplotype H), and an inverse association with Native American ancestry. An interesting finding revealed by a GWAS performed in Latin American women, showed a SNP in the ESR1 (estrogen receptor α gene) rs140068132, conferring protection against breast cancer. This SNP is present in 5%–15% of Latin American women depending on the Amerindian ancestry component. Carriers of this SNP present an OR = 0.6 (95% CI 0.53–0.67; $p < 9 \times 10^{-18}$) (Fejerman et al., 2014).

Our group analyzed Amerindian mitochondrial DNA haplogroups in a sample of 77 Chilean women with breast cancer and found a significant lower percentage of the matrilineal Amerindian component (69%) compared with the admixed general population in Chile (84%) (Rocco et al., 2002; Gallardo et al., 2006). Even when mitochondrial DNA markers represent only a matrilineal ancestry, this result is consistent with those mentioned earlier. In relation to AIMs no breast cancer studies have been published as such, but this is in need of further study. In Chile, the genetic structure of our population has been studied by "Chilegenómico" and others (http://www.chilegenomico.cl) (Ruiz-Linares et al., 2014; Eyheramendy et al., 2015). These results revealed that the Chilean population is on average 51% European, 5% African, and 44% Amerindian, with no relevant variations in percentages along Chile.

## Mutational Screening of *BRCA1* and *BRCA2* in Latin American Populations

The first full mutational screening on *BRCA1* and *BRCA2* in a cohort of hereditary breast cancer patients (56 Chileans) from Latin America was published by our group in 2006 (Gallardo et al., 2006). The selection criteria used in this study included patients belonging to families with at least (1) three relatives presenting breast cancer at any age, (2) two relatives with breast cancer with one having been diagnosed before the age of 43, (3) one breast and one ovarian cancer at any age, (4) one male and one female with breast cancer. We found that 20% (11/54) of nonrelated patients presented a mutation in *BRCA1* or *BRCA2*, with a total of seven mutations. Table 5.1 shows all reports including patients with family history including similar selection criteria in order to assemble genetic information on hereditary breast cancer in Latin America. These studies revealed percentages for BRCA mutation carriers between 13.7% and 26.3%, in Chile, Argentina, Colombia, Uruguay and Venezuela (Gallardo et al., 2006; Jara et al., 2006; Torres et al., 2007; Delgado et al., 2011; Lara et al., 2012; Solano et al., 2012). Among the eight studies on *BRCA1* and *BRCA2* mutational analysis and including familial cases of breast cancer, sample sizes range from 16 to 97. The other three studies from Argentina, Brazil and Cuba (Rodriguez et al., 2008; Solano et al., 2012; Carraro et al., 2013) have selected breast cancer patients who include a criteria recommended by the National Comprehensive Cancer Network (NCCN), such as patients with no apparent family history who present early onset breast cancer. As seen in Table 5.2 the number of included patients, screening techniques, and mutation rates vary widely among these studies. Considering Argentinanian and Brazilian studies carried on with DNA sequencing, the percentages of mutation carriers varied between 7.9% and 13.5% (Solano et al., 2012; Carraro et al., 2013). In Chile, we found that 11% of breast cancer patients with no apparent family history and early-onset (<40), carry a mutation in *BRCA1* or *BRCA2* (Alvarez et al., 2017). The study of 247 similar patients from Cuba (Rodriguez et al., 2008) revealed only 0.8% of mutation carriers, which may be due to the screening technique utilized or a real situation in this specific population. In Table 5.3 we gathered all publications screening all exons and intron-exon boundaries of *BRCA1* and *BRCA2*, but including patients with mixed selection criteria either with or without family history. As is shown, an additional Chilean study of 326 patients (Gonzalez-Hormazabal et al., 2011) indicates a 7% of mutation carriers, lower than the percentages found in previous studies in Chilean families selected by hereditary criteria. This result reveals the importance of selection criteria in these sort of studies. An NGS screening of 39 Mexican patients with breast cancer showed a 10.2% of mutation carriers (Vaca-Paniagua et al., 2012). The two most recent studies published in South America report a screening in more than 340 patients using Next Generation Sequencing (Fernandes et al., 2016;

**Table 5.2** *BRCA1* and *BRCA2* Mutation Screening in Latin American Breast Cancer Cases With No Family History

| Study Population | Number of Patients | Personal Breast Cancer Features | Screening Method | Patients with Mutation (%) | | Number of Mutations | | Number of Recurrent Mutations (% of total carriers) | Rearrangements |
|---|---|---|---|---|---|---|---|---|---|
| | | | | BRCA1 | BRCA2 | Total | Novel | | |
| Argentina Solano et al. (2012) | 37 | Single premenopausal women with breast cancer | Sanger Sequencing | 4/37 (10.8%) | 1/37 (2.7%) | 5 | 2 | 0 | ND |
| Brazil Carraro et al. (2013) | 38 | Single premenopausal women with breast cancer | Sanger Sequencing | 1/38 (2.6%) | 2/38 (5.3%) | 3 | 1 | 0 | ND |
| Cuba Rodriguez et al. (2008) | 247 | Single premenopausal women with breast cancer | PTT, DGGE | 1/247 (0.4%) | 1/247 (0.4%) | 2 | 0 | 0 | ND |

*Number of patients and mutation frequencies are referred only to no familial cases in each manuscript.*
*CSGE, Conformation sensitive gel electrophoresis; DGGE, denaturing gradient gel electrophoresis; DHPLC, denaturing high performance liquid chromatography; ND, not determined; NGS, next generation sequencing; PTT, protein truncation test.*

**Table 5.3** Other Studies in Latin American Breast Cancer Patients

| Study Population | Number of Patients | Screening Method | Families With Mutation (%) | | Number of Mutations | | Number of Recurrent Mutations (% of total carriers) | Rearrangements |
|---|---|---|---|---|---|---|---|---|
| | | | BRCA1 | BRCA2 | Total | Novel | | |
| Argentina[a] Vaca-Paniagua et al. (2012) | 940 | Sanger Sequencing, NGS, MLPA | 105/940 (11.1%) | 74/940 (7.9%) | 58 | 8 | 15(61%) | 5/940 |
| Brazil Solano et al. (2016) | 349 | Sanger Sequencing, NGS, MLPA | 49/349 (14%) | 26/349 (7.4%) | 43 | 6 | 8 (52%) | 2/349 |
| Brazil Villarreal-Garza et al. (2015a) | 145 | MLPA | ND | ND | ND | ND | ND | 5/145 |
| Chile[b] Alvarez et al. (2017) | 326 | CSGE, MLPA | 11/326 (3.3%) | 12/326 (3.7%) | 14 | 4 | 6 (65.2%) | 0 |
| Cuba Rodriguez et al. (2008) | 307 | PTT, DGGE | 1/307 (0.3%) | 7/307 (2.3%) | 6 | 0 | 1 (25%) | ND |
| Mexico Gonzalez-Hormazabal et al. (2011) | 39 | NGS | 2/39 (5.1%) | 2/39 (5.1%) | 4 | 3 | 0 | ND |

CSGE, Conformation sensitive gel electrophoresis; DGGE, denaturing gradient gel electrophoresis; DHPLC, denaturing high-performance liquid chromatography; HDA, heteroduplex analysis; HRM, high-resolution melting; MLPA, multiple ligation probe amplification; ND, not determined; NGS, next-generation sequencing; PTT, protein truncation test; SSCP, single-stranded conformational polymorphism.
aIncludes results from Solano et al. (2012).
bIncludes results from Jara et al. (2006).

Solano et al., 2016). These studies have included patients with a family history of breast and ovarian cancer as well as isolated patients. Among these studies mutation prevalence is 19% (Argentina) and 21.4% (Brazil) for breast and or ovarian cancer patients, respectively, selected by diverse criteria.

In terms of gene rearrangements described for *BRCA1* and *BRCA2* they seem to be scarce in Latin American populations. Only four countries, Argentina, Brazil, Colombia and Mexico, have reported pathogenic gene deletions and one Alu insertion (Villarreal-Garza et al., 2015a; Ewald et al., 2016; Fernandes et al., 2016; Solano et al., 2016) and one report from Chile presented non-pathogenic amplifications in *BRCA1* in three patients (Sanchez et al., 2011). In Argentina three breast and two ovary cancer patients presented a deletion in *BRCA1*, constituting 0.54% of total patients (Solano et al., 2016). In Brazil two rearrangements were found in two patients (Fernandes et al., 2016), one of those c.156_157insAlu being founder in Portugal (reviewed in Ossa and Torres, 2016) and also described in a second Brazilian study in three patients (Ewald et al., 2016). The two studies from Brazil give percentages of breast cancer patients carrying a rearrangement between 0.57% and 3.4% (Ewald et al., 2016; Fernandes et al., 2016). A study in Mexican patients with breast and ovarian cancer, unselected for family history (Villarreal-Garza et al., 2015a) a founder deletion in *BRCA1*, ex9-12del, previously described in Hispanic from Mexican origin (Weitzel et al., 2007) was detected in 33% of mutation carriers (6.9% of total patients).

## Founder Mutations in Latin American Breast Cancer Families

In Latin America, there are a few different mutations with a demonstrated founder effect that have been described in different populations: Mexico (*BRCA1* del exons 9–12), Brazil (*BRCA1* 5382insC), Colombia (*BRCA1* 3450delCAAG, A1708E, and *BRCA2* 3034del4) and US Hispanics (*BRCA1* 185delAG, IVS5+1G>A, S955X, and R1443X) (reviewed in Ashton-Prolla and Vargas, 2014; Ossa and Torres, 2016). In addition to these founder mutations, other recurrent mutations have been described in Latin American populations in which a founder effect has still not been tested. Another relevant finding is that founder and recurrent mutations have accounted for a high percentage of *BRCA1* and 2 mutation carriers, which have provided a great opportunity for setting up a simple screening in breast cancer patients. So far, there have been three examples in Latin American populations published that describe this event. In Colombia a specific study done in 53 breast/ovarian cancer families from Bogotá revealed the presence of three founder mutations accounting for 80% of *BRCA1/2* mutation carriers (Torres et al., 2007). A study from Mexico (Torres-Mejia et al., 2015) of 810 patients from three different cities showed that six recurrent mutations account for 62% of total carriers. And, in Brazil an extended study of 349 breast cancer patients described that the founder mutation 5382insC (now c.5266insC) represents 24% of all carrier families and 11% presented 3450delCAAG (now c.3331_3334delCAAG) (Fernandes et al., 2016). In Chile, our group has already screened *BRCA1* and *BRCA2* in 453 patients with breast cancer patients, with or without family history. The results of this study, in addition to previous ones, led us to define nine founder mutations in Chilean patients (Alvarez et al., 2017). The most striking finding is that these nine founder mutations are present in 78% of the 71 Chilean breast cancer carriers detected among the 453 patients. In a collaborative study, we have demonstrated that mutation 3450delCAAG, which is also founder in Colombia, has a common Spanish origin that arrived independently in Chile and Colombia (Tuazon et al., manuscript in preparation). It is worth pointing out that, except for the Ashkenazi Jewish mutations and the highly recurrent 3450delCAAG in *BRCA1*, the recurrent mutations found in Latin America do not overlap between the different countries on the continent. These findings strongly support the idea that a specific panel of recurrent and or founder mutations must be generated independently in each Latin American country for the screening of *BRCA1* and *BRCA2* mutations in breast cancer patients.

## Shared Mutations Among Latin American Breast Cancer Patients

Among their patients, Brazil and Argentina show the highest number of shared mutations in *BRCA1* and *BRCA2* (*n* = 13), followed by Mexico/Argentina (*n* = 8), Chile/Argentina (*n* = 7) and Chile/Brazil (*n* = 6). The other countries scarcely share one or two mutations with other Latin American populations (Colombia, Costa Rica, Peru, Puerto Rico, Uruguay and Venezuela), or none (Cuba). Observed in more countries is the mutation *BRCA2* c.2808_2811delACAA (also described as c.2806_2809delAAAC, 3034del4, or 3036del4), which has been described recurrently in Europe and is a founder mutation in Colombia (Torres et al., 2007). This mutation is recurrent in Argentina (Solano et al., 2016), Brazil (Fernandes et al., 2016), Colombia (Torres et al., 2007) and Peru (Abugattas et al., 2015), reported in two to eight patients, and observed only once in Venezuela (Lara et al., 2012) and Mexico (Torres-Mejia et al., 2015). Another mutation observed in several countries is the Spanish founder mutation c.211A>G (also known as R71G), found in one patient from Chile (Alvarez et al., 2017) and Mexico (Villarreal-Garza et al., 2015a), and recurrent in Argentina (*n* = 11) (Solano et al., 2016) and in the northeast of Brazil (*n* = 5) (Felix et al., 2014). Another 31 mutations are shared only by two or three countries and close to a hundred are not even shared, so the vast majority of mutations in Latin American patients are not common.

A different approach for *BRCA1* and *BRCA2* screening has been carried out by the group of J. Weitzel who defined a panel of 50 *BRCA1* and 46 *BRCA2* mutations recurrent in Hispanic breast cancer patients from the United States, Mexico, and Colombia, as other mutations taken from publications (Weitzel et al., 2013) (patent application US 20130183667A1). This panel was applied to 280 breast cancer patients at a hospital in Medellin (Colombia) unselected for family history, finding only 1.2% of mutation carriers (Hernández et al., 2014). A similar study was performed in 266 patients from a public Hospital in Lima, finding 5% of mutation carriers (Abugattas et al., 2015), among which the Ashkenazi Jewish mutation 185_186delAG was present in 7 of 13 patients with mutations. The main issue with this panel is, as has been noticed already (Ashton-Prolla and Vargas, 2014), that recurrent mutations among Latin American countries are different, so a panel trying to gather all these mutations will not be successful.

Ashkenazi Jewish founder mutations (185delAG, 5382insC in *BRCA1* and 6174delT in *BRCA2*) have been observed in several Latin American countries, although in Brazil and Argentina they are highly recurrent. Mutation 5382insC, which is not exclusively Ashkenazi Jewish (Hamel et al., 2011), was described in 2007 in Brazilian patients. Posterior screening of a wider Brazilian population, found 18% and 24% of patients with the 5382insC mutation (Carraro et al., 2013; Fernandes et al., 2016). In Argentina the three mutations were represented, being 185delAG being the most recurrent with 13.4%, 6174delT with 11.7% and 5382insC with 7.8% (Solano et al., 2016). Together the three mutations account for 33% of all mutations carriers in Argentina (Solano et al., 2016). In Chile, considering all studies, only 5.3% of patients carry one of the three Jewish Founder mutations. In Costa Rica, only mutation 6174delT has been identified but in one patient (Gutiérrez Espeleta et al., 2012). Mutation 185delAG has been found in Mexico and Peru through HISPANEL analysis with frequencies 2.5%–6.8% in Mexico (Villarreal-Garza et al., 2015a,b) and 53.8% in Peru (Abugattas et al., 2015).

# LYNCH SYNDROME

## Clinical and Genetic Features

In 1984, the term Lynch syndrome was proposed to describe families with colorectal and gynecological cancer with a lack of polyposis phenotype, and this term has been commonly used since then (Boland and Troncale, 1984). This Syndrome is the most common inheritable type of colorectal cancer, accounting for 2%–4% of the cases, and has an estimated prevalence in the general population of 1 in 440 (Hampel et al., 2005; Hampel et al., 2006; Rubenstein et al., 2015). Although the term hereditary nonpolyposis colorectal cancer (HNPCC) is often used interchangeably with Lynch syndrome, it is important to remember that HNPCC is a clinical diagnosis for patients and/or families that meet Amsterdam I or II criteria, whereas the diagnosis of Lynch syndrome requires the presence of a genetically confirmed pathogenic mismatch repair (MMR) variant (Sjursen et al., 2010; Dominguez-Valentin et al., 2015; Da Silva et al., 2016; Tiwari et al., 2016). To select families for genetic analysis, the Amsterdam criteria were developed to permit the identification of MMR defects and their association within tumor spectrum (Vasen et al., 1999; Umar et al., 2004). The Amsterdam criteria requires at least three affected family members in two or more generations, with one being a first-degree relative of the other two and at least one individual diagnosed before 50 years of age (Vasen and Muller, 1991; Vasen et al., 1999). Amsterdam I applies to families with three or more cases of colorectal cancer whereas Amsterdam-II also includes extracolonic tumors, that is, endometrial cancer, cancer of the upper urinary tract and cancer of the small bowel (Vasen and Muller, 1991; Vasen et al., 1999). The Bethesda guidelines, originally designed to identify tumors likely to have microsatellite instability (MSI), include aspects of the tumor phenotype, but are less strict with regard to family history (Rodriguez-Bigas et al., 1997; Umar et al., 2004). For this reason, families defined by the Amsterdam criteria have shown a higher incidence of MMR gene mutations than those defined by Bethesda. On the other hand, MSI or immunohistochemical testing of tumors are used to select patients with a new colorectal cancer diagnosis derived from germline diagnostic testing (Berg et al., 2009), Lynch syndrome patients have an increased risk for the following cancers: colorectal (lifetime risk = 70%–80%), endometrial (50%–60%), stomach cancer (13%–19%), ovarian cancer (9%–14%), small intestine, biliary tract, and brain as well as carcinoma of the ureters and renal pelvis (Kobayashi et al., 2013).

Lynch syndrome is caused by germline pathogenic variants in one of the MMR genes: *MLH1*, *MSH2*, *MSH6* and *PMS2* or deletion in the *EPCAM* gene, which leads to methylation of the adjacent *MSH2* promoter. Such variants are here referred to as path_MMR (pathogenic variants in mismatch repair genes) and, when specifying one of the genes, as *path_MLH1*, *path_MSH2*, *path_MSH6*, *path_PMS2* or *path_EPCAM* (Møller et al., 2015). The cumulative incidence of any cancer at 70 years of age is 72% for *path_MLH1* and *path_MSH2* carriers but lower in *path_MSH6* (52%) and *path_PMS2* (18%) carriers. *Path_MSH6* and *path_PMS2* carriers do not have increased risk for cancer before 40 years of age (Møller et al., 2015; Møller et al., 2016). To date over 3100 unique DNA variants across the MMR genes have been described in the International Society for Gastrointestinal Hereditary Tumors (InSIGHT) (http://insight-group.org/variants/database/), and a recent clinical InSiGHT consensus classification has identified 57% of the MMR variants as pathogenic or likely pathogenic (Class 5 and 4), 32% as uncertain variants (Class 3), 4% as likely not pathogenic (Class 2) and 7% as not pathogenic (Class 1) (Thompson et al., 2014; Da Silva et al., 2016).

## Hereditary Lynch Syndrome Registries in Latin America

According to the Study Group on Hereditary Tumors (GETH, http://www.geth.org.br/novo/) and a recent thorough investigation of most of the Latin American countries, there are 32 hereditary cancer care centers in Latin America: 5 in Argentina, 1 in Bolivia, 20 in Brazil, 1 in Chile, 1 in Colombia, 1 in Mexico, 1 in Peru, 1 in Puerto Rico and 1 in Uruguay (Vaccaro et al., 2016; Rossi et al., 2017). These registries differ in fundamental aspects of function, capabilities and funding, but are able to conduct high quality clinical, research and educational activities because of the dedication and personal effort of their members, and organizational support (Cruz-Correa et al., 2015; Vaccaro et al., 2016).

Limitation on genetic testing has an impact in the evaluation of patients at risk for hereditary cancer and their relatives, and ultimately increases the burden of cancer for this minority population (Cruz-Correa et al., 2015). In Latin America, genetic testing is not routinely available in the public health system, with the exception of a few studies conducted in research institutes or private institutions. For instance, until recently the coverage of oncogenetic services, that is, in Brazil, was restricted to less than 5% of the population. However, a significant advance took place in 2012, when the coverage of genetic testing by private health care plans became mandatory in Brazil, currently covering around 20%–30% of the population (Viana et al., 2008; Ashton-Prolla and Seuanez, 2016). In Puerto Rico, insurance coverage for genetic testing in hereditary cancer risk assessment has been included in the Affordable Care Act ; however, the genetic testing for Lynch syndrome is not directly mandated by the Act (Cruz-Correa et al., 2015). In this regard, Medicare covers genetic testing for Lynch syndrome for individuals who meet AMS criteria or Bethesda guidelines, per the National Comprehensive Cancer Network (NCCN) (Cruz-Correa et al., 2015).

## Mutation, Molecular and Clinical Profile of Lynch Syndrome in Latin America

We recently characterized the clinical, molecular and MMR variants spectrum of families from eleven Lynch syndrome hereditary cancer registries and published databases from Latin America (Roque et al., 2000; Rossi et al., 2002; Giraldo et al., 2005; Sarroca et al., 2005; Chialina et al., 2006; Clarizia et al., 2006; Montenegro et al., 2006; Vaccaro et al., 2007a,b; Viana et al., 2008; Alvarez et al., 2010; De Jesus-Monge et al., 2010; Leite et al., 2010; Ricker et al., 2010a; Alonso-Espinaco et al., 2011; Egoavil et al., 2011; Koehler-Santos et al., 2011; Valentin et al., 2011; Ramírez-Ramírez et al., 2012; Rasuck et al., 2012; Santos et al., 2012; Valentin et al., 2012; Wielandt et al., 2012; Dominguez-Valentin et al., 2013; Castro-Mujica et al., 2014; Marqués-Lespier et al., 2014; Nique Carbajal et al., 2014; Carneiro da Silva et al., 2015; Cruz-Correa et al., 2015;

de Freitas et al., 2015; Garcia et al., 2015; Cajal et al., 2016; Castro-Mujica et al., 2016; Dominguez-Valentin et al., 2016; Germini et al., 2016; Moreno-Ortiz et al., 2016; Ortiz et al., 2016; Rossi et al., 2016; Vaccaro et al., 2016; Rossi et al., 2017). We identified 6 out of 15 countries, that is, Argentina, Brazil, Chile, Colombia, Uruguay and Puerto Rico, where germline genetic testing for Lynch syndrome is available and 3 countries, that is, Bolivia, Peru, and Mexico, where tumor testing is used in the diagnosis of Lynch syndrome (Rossi et al., 2017). The spectrum of pathogenic MMR variants included *path_MLH1* up to 54%, *path_MSH2* up to 43%, *path_MSH6* up to 10%, *path_PMS2* up to 3% and *path_EPCAM* up to 0.8%. Frequent regions included exon 11 of *MLH1* (15%), exons 3 and 7 of *MSH2* (17 and 15%, respectively), exon 4 of *MSH6* (65%), exons 11 and 13 of *PMS2* (31% and 23%, respectively) (Rossi et al., 2017).

Fifteen published data from Latin America MMR Lynch syndrome spectrum contained information for 982 suspected Lynch syndrome families (Table 5.4). Overall, considering all studies, 37.9% of patients have a mutation but the higher mutation rate is obtained in patients meeting Amsterdam criteria (204/355, 57.4%), compared with Bethesda patients (70/383, 18.2%). From the 373 path_MMR carriers, *MLH1* was affected in 52.5% (196/373), *MSH2* in42.4% (158/373), *MSH6* in 3.8% (14/373), *PMS2* in 0.8% (3/373) and *EPCAM* in 0.5% (2/373) (Table 5.4) (Rossi et al., 2017). Seven founder mutations have also been identified among the Latin American Lynch syndrome families. Four of them have been suggested to constitute potential founder mutations in other populations, for example, the Italian-Quebec *MLH1* c.545+3A>G, the Newfoundland *MSH2* c.942+A>T, the Portuguese *MSH2* c.388_389delCA and the Spanish *MSH2* exon 7 deletion (Hampel et al., 2005; Kobayashi et al., 2013; Møller et al., 2015; Rubenstein et al., 2015). In Colombia, the *MSH2* c.1039-8T_1558+896dup was suggested to represent a founder mutation and in Chile, the *MLH1* c.1731+3A>T and the *MSH2* c.2185_2192delATGTTGGAinsCCCT were associated with a strong Amerindian genetic ancestry (Alvarez et al., 2010; Alonso-Espinaco et al., 2011; Dominguez-Valentin et al., 2013; Vaccaro et al., 2016). The Latin American population consists mainly of an admixture of Amerindians and Europeans, with a lower component of African ancestry. The proportion of these ancestries is variable among countries, being Argentina, Brazil, and Uruguay having a stronger European component. In Chile the European and Amerindian components are 45% and 55% respectively; in Colombia, Peru and Bolivia, a large part of the populations has Spanish colonist and American Indian ancestry, and in Brazil a significant part of the population has African and American Indian roots (Dominguez-Valentin et al., 2013; Vaccaro et al., 2016).

Interestingly, the clinic pathological features of pathogenic MMR carriers described in Latin American families are in accordance with other studies,

**Table 5.4** Summary of Published Data From Latin America Lynch Syndrome Spectrum

| Study Population | N (families/ patients) | Selection Criteria | | | Families With Mutation | | Path_MLH1 Carriers (%) | Path_MSH2 Carriers (%) | Path_MSH6 Carriers (%) | Path_PMS2 Carriers (%) | Path_EPCAM Carriers (%) |
|---|---|---|---|---|---|---|---|---|---|---|---|
| | | AMS | Revised Bethesda | Other Criteria | AMS | Revised Bethesda | | | | | |
| Argentina Viana et al. (2008) | 1 | 1 | NA | NA | 1 | NA | 0 | 1 (100) | NA | NA | NA |
| Brazil Ashton-Prolla and Seuanez (2016) | 25 | 6 | 18 | 1 | 3 | 7 | 8(80) | 2(20) | NA | NA | NA |
| Uruguay Rossi et al. (2002) | 12 | 12 | NA | NA | 3 | NA | 2(67) | 1 (33) | 0 | NA | NA |
| Colombia Roque et al. (2000) | 23 | 11 | 12 | NA | 9 | 2 | 10(91) | 1 (9) | NA | NA | NA |
| Argentina Clarizia et al. (2006) | 11 | 11 | NA | NA | 5 | NA | 2(40) | 3 (60) | NA | NA | NA |
| Mexico, El Salvador and Guatemala Ortiz et al. (2016) | 12 | 5 | 7 | NA | 5 | 6 | 7(64) | 4 (36) | NA | NA | NA |
| Chile Vaccaro et al. (2007a) | 21 | 14 | 7 | NA | 7 | 2 | 6(67) | 3(33) | NA | NA | NA |
| Colombia De Jesus-Monge et al. (2010) | 1 | 1 | NA | NA | 1 | NA | 1(100) | 0 | NA | NA | NA |
| Southeastern Brazil, Buenos Aires and Montevideo Egoavil et al. (2011) | 123 | 57 | 66 | NA | 25 | 9 | 20(59) | 14(41) | NA | NA | NA |
| Chile Santos et al. (2012) | 35 | 19 | 16 | NA | 18 | 3 | 14(67) | 5(24) | 2(9) | NA | NA |
| South America Valentin et al. (2012) | 267 | 147 | 120 | NA | 81 | 18 | 59(60) | 40(40) | NA | NA | NA |
| Puerto Rico and Dominican Republic Vaccaro et al. (2016) | 89 | 19 | 70 | NA | 12 | 10 | 8(36) | 13(59) | 1(5) | NA | NA |
| Southeastern Brazil Marqués- Lespier et al. (2014) | 116 | 49 | 67 | NA | 31 | 13 | 15(33) | 25(56) | 4(9) | 1(2) | NA |
| Mexico Dominguez-Valentin et al. (2016) | 3 | 3 | NA | NA | 3 | NA | 2(67) | 1(33) | NA | NA | NA |
| South America Thompson et al. (2014) | 243 | NA | NA | NA | NA | NA | 42(43) | 45(46) | 7(7) | 2(2) | 2(2) |
| Total | 982 | 355 | 383 | 1 | 204 | 70 | 196 (52.5) | 158 (42.4) | 14 (3.8) | 3 (0.8) | 2 (0.5) |

AMS, Amsterdam criteria; MMR, mismatch repair; N, number; path_MLH1, pathogenic (disease-causing) variant of the MLH1 gene; Path_MMR, pathogenic (disease-causing) variant of an MMR gene; path_MSH2, pathogenic (disease-causing) variant of the MSH2 gene; path_MSH6, pathogenic (disease-causing) variant of the MSH6 gene; path_PMS2, pathogenic (disease-causing) variant of the PMS2 gene.
Source: Modified from Rossi et al. 2017.

that is, the Amsterdam II criteria were fulfilled by 64% of *path*_MMR carriers, although MSI, IHC, and family history are still the primary criteria in several countries, where no genetic testing for Lynch syndrome is available yet (Rossi et al., 2017). However, awareness of hereditary cancer among clinicians involved in the diagnosis and treatment of colorectal cancer is currently low, but families actually meeting the clinical criteria may not have been identified (Sjursen et al., 2016). In addition, the average life expectancy in Latin America and the Caribbean is 75 years and inequalities persist among and within the countries (www.paho.org). These countries are mainly represented by a young population where family history could be less informative and not sensitive to assessing genetic screening for Lynch syndrome (Rossi et al., 2017).

In summary, the Latin American Lynch syndrome mutation spectrum includes multiple new mutations, international founder effects, genetically frequent regions, and potential founder mutations, which will be useful for future the development of genetic testing.

## Latin American Collaborative Network

In Latin America, the application of genetic tests to patients as part of clinical practice is a big challenge because of low budgets in public health. In this sense, international collaborations would be helpful for the development of additional studies on Lynch syndrome in Latin American countries to both, increase the knowledge of MMR variants in different populations and to bring additional awareness of this condition to medical professionals and public health leaders in Latin America.

# FAMILIAL ADENOMATOUS POLYPOSIS

## Clinical and Genetic Features

Familial adenomatous polyposis (FAP, OMIM No. 175100) is an autosomal dominant disease with complete penetrance, which account for about 1% of colorectal cancers. Clinical manifestations include the development of hundreds to thousands of adenomatous polyps located mainly in the colon and rectum, and the development of colorectal cancer at 35–40 years of age (Strate and Syngal, 2005). Consequently, the recommended treatment for FAP patients is prophylactic colectomy in young adulthood. Moreover, individuals with FAP can present a number of benign extracolonic features, including multiple osteomas, epidermoid cysts, desmoid tumors, and congenital hypertrophy of the retinal pigment epithelium. According to the polyp number, the phenotype can be classified as classical (more than 100 polyps) or attenuated (fewer than 100 polyps) (Knudsen et al., 2003), this classification is very important in order to select families for genetic studies.

FAP is caused by mutations in the tumor suppressor gene APC (adenomatous polyposis coli gene), located in chromosome 5q21-22 (Leppert et al., 1987; Nakamura et al., 1988; Kinzler et al., 1991). The APC gene consists of 15 coding exons, and its cDNA spans more than 8529 bp (Groden et al., 1991). It encodes a protein whose main function is partially controlling the cell-cycle progression by participating in the Wnt signaling pathway. The APC protein regulates the degradation of β-catenin, a transcriptional activator of genes, such as cyclin D1 and c-myc proto-oncogenes (He et al., 1998).

To date more than 1700 different APC genetic alterations have been reported in The Human Gene Mutation Database (HGMD; http://www.hgmd.cf.ac.uk/) and more than 1190 genetic alterations have been described in the Leiden Open Variation Database (LOVD; http://www.lovd.nl/APC). Most of them correspond to nonsense (24%) and frameshift (59%) type, resulting in truncated proteins. Different studies in patients with FAP have shown a high correlation between the site of the mutation and the clinical phenotype, such as grade of severity and/or the presence of extracolonic manifestations (Fearnhead et al., 2001; Merg et al., 2005; Galiatsatos and Foulkes, 2006). Mutations located between codons 1255 and 1467 are associated with the presence of more than 1000 polyps in the colon and rectum (Nagase et al., 1992). However, mutations in the 5′ and 3′ ends of the gene have been associated with attenuated FAP with less than 100 polyps (Spirio et al., 1993; Friedl et al., 1996). Conversely, diverse extracolonic features, such as osteomas, congenital hypertrophy of retinal pigment epithelium (CHRPE), and desmoid tumors, correlate with mutations in a specific region of the APC gene between codons 767 and 1513, 457 and 1444, and 1310 and 2011, respectively (Caspari et al., 1995; Bertario et al., 2003; Bisgaard and Bülow, 2006). Among all mutations described, the most frequent are c.3183_3187delACAAA and c.3927_3931delAAAGA, also called hot spots 1061 and 1309, respectively. Patients carrying the latter mutation have been found in different populations and present a profuse polyposis and early-onset age of the disease (Caspari et al., 1994).

## Mutational Screening of APC Gene in Latin America

To date five Latin American countries have reported studies on the mutational screening of the APC gene in FAP patients: Argentina, Brazil, Chile, Cuba Puerto Rico (Cruz-Bustillo et al., 2002; De Rosa et al., 2004; De La Fuente et al., 2007; Cruz-Correa et al., 2013; De Queiroz Rossanese et al., 2013; Torrezan et al., 2013). One additional report showed the findings for nine Hispanic patients from México, Guatemala, and Honduras (Ricker et al., 2010b). As shown in Table 5.5, a total of 163 patients have been studied, whose main criterion of selection was classic FAP ($n = 116$) and a lower proportion of patients with attenuated phenotype ($n = 10$) and with an unknown number of polyps ($n = 37$). The percentage of patients with mutations varies significantly among

**Table 5.5** Summary of *APC* Mutations in Latin American Studies

| Study Population | N (families) | Selection criteria[a] | Technique | % Families with mutation | Mutations Total | Mutations Novel | Common mutation in codon 1309 | Exon 15 % mutations | Exon 15 Range |
|---|---|---|---|---|---|---|---|---|---|
| Argentina Bertario et al. (2003) | 51 | 49 classical; 2 attenuated | PTT/SSCP/qPCR | 39/51 (77%) (37 classical and 2 attenuated) | 29 | 13 (45%) | 10/39 (26%) | 26/29 (90%) | c.2522-c.4891 |
| Brazil Caspari et al. (1994) | 23 | 15 classical; 8 attenuated | Sanger sequencing/MLPA/ArrayCGH/Duplex-qPCR | 15/23 (65%) (13 classical and 2 attenuated) | 14 | 5 (36%) | 2/15 (13%) | 8/14 (57%) | c.3050-c.5365 |
| Brazil De Rosa et al. (2004) | 20 | 20 classical | Sanger sequencing | 15/20 (75%) | 13 | ND | 1/15 (7%) | 13/13 (100%) | c.2628-c.4692 |
| Chile Torrezan et al. (2013) | 24 | 24 classical | SSCP/PTT | 21/24 (88%) | 17 | 9 (53%) | 3/21 (14%) | 15/17 (88%) | c.2486-c.4709 |
| Cuba De Queiroz Rossanese et al. (2013) | 17 | Polyposis # unknown | HAD (Partial study of exon 15) | 5/17 (29%) | 3 | 0 | 3/5 (60%) | 3/3 (100%) | c.3183-c.3925 |
| Hispanics Cruz-Bustillo et al. (2002) | 9 | 8 classical; 1 polyposis # unknown | Sanger sequencing/Rearrangements study | 7/9 (78%) (6 classical and 1 unknown) | 7 | ND | 1/7 (14%) | 4/7 (57%) | c.3183-c.3709 |
| Puerto Rico De La Fuente et al. (2007) | 19 | Polyposis # unknown | Commercial sequence analyses | 8/19 (42%) | 8 | ND | 1/8 (13%) | 6/8 (75%) | c.3149-c.4612 |
| TOTAL | 163 | 116 classical; 10 attenuated; 37 unknown | | 110/163 (67%) 92/116 (79%) 4/10 (40%) | 91 | | 21/110 (19%) | 75/91 (82%) | c.2486-c.5365 (codon 829-1788) |

Array-CGH, Comparative genomic hybridization on microarray; HDA, heteroduplex analysis; MLPA, multiple ligation probe amplification; ND, not determined; NGS, next generation sequencing; PTT, protein truncation test; qPCR, quantitative PCR; SSCP, single-stranded conformational polymorphism.
aA classical >100 polyps and attenuated <100 polyps.

these studies (29%–88%), which may be explained by variable selection criteria of patients and technical approaches used for mutation detection. For example, Cuba reported the lowest percentage of mutations (29%) since their study was done by heteroduplex technique. On the contrary, when a combination of SSCP, PTT and Sanger sequencing was used in classic FAP patients the percentage of mutation carriers rose to 88% (Table 5.5). On average, considering all studies, 67% of patients have a mutation in APC, but the higher mutation rate is obtained in patients showing a classic FAP phenotype (79%), compared to patients with attenuated FAP (40%). Mutations were localized mainly in exon 15 of *APC* (82%), specifically between cDNA nucleotides 2486 and 5365 (codon 829 and 1789). This part of the gene includes the "mutational cluster region" described for APC, which includes codon 1309, already identified as mutated in 19% of Latin American patients. New mutations have been reported in three publications representing between 36% and 53% of all mutations detected. The significant rate of novel mutations reported between studies may be attributed to the scarce information in databases regarding Latin American patients.

In summary, 76 different mutations have presently been reported until today for the Latin American population, the majority being point mutations (De Jesus-Monge et al., 2010), three gene rearrangements involving the deletion of the APC gene or its partial duplication. For this reason DNA sequencing is highly effective in the screening of mutations in the FAP patients since only few patients will need to be studied for gene rearrangements. Finally, seven mutations are shared between different Latin American countries, two of which are localized in the hot spots already defined, and present in patients from all Latin American countries. Among shared mutations we found the following: c.2626C>T, c.4348 C>T and c.4393_4394delAG present in Brazil and Argentina; c.3783_3784delTT and c.4280delC present in Argentina and Chile; the hot spot c.3183_3187delACAAA (codon 1061) identified in Argentina, Brazil, Cuba, Mexico and Puerto Rico; and the hot spot c.3927_3931delAAAGA (codon 1309) present in Argentina, Brazil, Chile, Cuba, Mexico and Puerto Rico.

## *MUTYH* Associated Polyposis

In 2002, Al-Tassan et al. (2002) determined that some FAP patients with no mutation in the APC gene have germinal mutation in the MUTYH gene. This type of polyposis, called *MUTYH Associated Polyposis* (MAP), was characterized as an autosomic recessive disease with an indistinguishable phenotype from FAP, which is caused by mutations in the APC gene (Al-Tassan et al., 2002; Sampson et al., 2003). The MUTYH gene codes for a DNA glycosylase enzyme that repairs oxidative DNA damages caused by ROS (Ohtsubo, 2000; Cheadle and Sampson, 2003). The more stable oxidized guanine

(7,8-dihydroxy-8-oxoguanine, also named 8-oxo-G) pairs adenine instead of cytosine during DNA replication leading into a genetic transvertion of guanine:citosine for timine:adenine (G:C > T:A). Normally, MUTYH excises these adenines mispaired with oxidized guanine (Oka and Nakabeppu, 2011). Hence, the absence of MUTYH function leads to an increase somatic mutations, in target genes as APC and in addition an escape from programed cell death permitting tumor development (Al-Tassan et al., 2002).

To date, 308 different MUTYH genetic alterations have been reported in the LOVD (http://www.lovd.nl/MUTYH). Most correspond to nucleotide substitutions (97.6%) and in less proportion genomic rearrangements (2.3%). Two genetic variants have been frequently reported in Caucasian populations: p.Y165C (c.494A>G) in exon 7 and p.G382D (c.1145G>A) in exon 13, 535 and 509 times, respectively. A meta-analysis for these two variants determined that *MUTYH* homozygous carriers demonstrated a 28 fold increased risk of colorectal cancer (95% CI: 6.95–115), and monoallelic carriers have a marginal effect with a OR = 1,34 (95% CI: 1.00–1.80) (Theodoratou et al., 2010).

### Mutational Screening of *MUTYH* in Latin America

In total, only four countries in Latin America have published studies on the mutational screening of *MUTYH*: Argentina, Brazil, Puerto Rico, and Chile (De Rosa et al., 2004; Álvarez et al., 2012; Cruz-Correa et al., 2013; Torrezan et al., 2013). The selection criteria included patients with FAP patients negative for APC mutations or attenuated FAP. As shown in Table 5.6, Caucasian mutations (Y165C and G382D) are highly frequent in Puerto Rican (10/13) and Brazilian (4/6) FAP patients (118,122). On the contrary, in Argentina no mutations have been identified (De Rosa et al., 2004), and in Chile the only mutation detected was c.340c>T/p.Y114H in homozygocity (unpublished results) (Álvarez et al., 2012).

## CONCLUSION

Mutational screening in cancer genes, more specifically in breast and colorectal cancer, has been extensively studied in Latin American populations in the last 10 years. Valuable information has been gathered on the genetics of these two types of cancer, which is compiled and discussed in this chapter. Even though cancer patients from some countries have not yet been tested, the information gathered to date from studies in populations across Latin America gives relevant genetic data. We have reliable estimates on the percentages of mutations carriers in genes of high risk for these two types of hereditary cancer and enough expertise in gene screening in several Latin American countries. On the basis of these genetic analyses the clinical management of patients has been improved and will surely derive result in better life expectancy for patients, now and in the future.

**Table 5.6** Summary of MUTYH Mutations in Latin American Studies

| Population | N (families) | Selection Criteria[a] | Technique | % Families With Mutation | Mutations | | Other Mutations |
|---|---|---|---|---|---|---|---|
| | | | | | G382D homozygous | G382D/ Y165C heterozygous | |
| Brazil Caspari et al. (1994) | 8 | 1 classical; 7 attenuated | Allele-specific PCR/MLPA/ Array-CGH/ Duplex-qPCR | 6/8 (75%) (1 classic and 5 attenuated) | 1 | 0 | Deletion exons 4–16 homozygous p.Y165C/p.A385Pfsª23 heterozygous p.Y165C/p.V130Efsª98 heterozygous p.R241W homozygous p.Y165C/p.E410Gfsª43 heterozygous |
| Puerto Rico De La Fuente et al. (2007) | 13 | Polyposis # unknown | Commercial sequence analyses | 13/13 (100%) | 6 | 3 | p.G382D monoallelic p.N42S monoallelic p.E465del monoallelic IVS7+8G>C monoallelic |
| Chile Oka and Nakabeppu (2011) | 1 | FAP classical negative-APC | SSCP | 1/1 (100%) | 0 | 0 | p.Y114H homozygous |
| TOTAL | 22 | 2 classical; 7 attenuated; 13 polyposis # unknown | | | 7 | 3 | 10 |

Array-CGH, Comparative genomic hybridization on microarray; MLPA, multiple ligation probe amplification; qPCR, quantitative PCR.
aA classical >100 polyps and attenuated <100 polyps.

Off course more research in cancer genetics needs to be done in Latin American countries, to cover other populations not studied until today, to extend the number of patients within each country, and to know the genetics of other hereditary types of cancer. In addition, ancestry determination in cancer patients is highly relevant in order to determine the influence of ethnic origins in the development of cancer. To increase knowledge and expertise in cancer genetics in Latin America, a significant increase in state investment is required in the funding of research grants, the training of young scientists in genetics, and the installation of PhD programs in genetics and genomics. It is not by chance that Brazil, the country with the highest investment in Science (1.24% GDP; 2013) in Latin America, has the most developed research in genetics and genomics: Brazil's expertise in this area is revealed in the number of papers, grants, infrastructure, and technology related to genetics and genomics. It is true that science in Latin America has become a more visible presence because of countries where science has become more developed. Those countries started establishing PhD programs 30 years ago and began developing research in their own laboratories with continued collaboration with European or North American research groups. The high qualification of scientists has raised the impact of research, with very small funds. The majority of Latin American countries allocate scarce investment to scientific research (Argentina: 0.88% GDP; Chile: 0.38% GDP; Mexico: 0.54% GDP); hence the number of scientists, research grants, young scientists, and technology is insufficient to cover the fields in need of investigation fields. Latin American countries need more investment in research, in education of young scientists, improvement of technology, but all these need to be developed in house, in Latin American Laboratories. Collaborations between our researchers and scientists from countries with a more developed science are very welcome and necessary; however if samples from our populations are only taken out for analysis, but not shared with our scientist, there will be no improvement in our research nor developments in our science.

## References

Abugattas, J., Llacuachaqui, M., Allende, Y.S., Velásquez, A.A., Velarde, R., Cotrina, J., et al., 2015. Prevalence of BRCA1 and BRCA2 mutations in unselected breast cancer patients from Peru. Clin. Genet. 88 (4), 371–375.

Alonso-Espinaco, V., Giráldez, M.D., Trujillo, C., van der Klift, H., Muñoz, J., Balaguer, F., et al., 2011. Novel MLH1 duplication identified in Colombian families with Lynch syndrome. Genet. Med. 13 (2), 155–160.

Al-Tassan, N., Chmiel, N.H., Maynard, J., Fleming, N., Livingston, A.L., Williams, G.T., et al., 2002. Inherited variants of MYH associated with somatic G:C-->T:A mutations in colorectal tumors. Nat. Genet. 30 (2), 227–232.

Alvarez, K., Hurtado, C., Hevia, M.A., Wielandt, A.M., De La Fuente, M., Church, J., et al., 2010. Spectrum of MLH1 and MSH2 mutations in Chilean Families with suspected Lynch syndrome. Dis. Colon Rectum. 53 (4), 450–459.

Alvarez, C., Tapia, T., Perez-Moreno, E., Gajardo-Meneses, P., Ruiz, C., Rios, M., et al., 2017. *BRCA1* and *BRCA2* founder mutations account for 78% of germline carriers among hereditary breast cancer families in Chile. Oncotarget 8 (43), 74233–74243.

Álvarez, K., de la Fuente, M., Orellana, P., Wielandt, A.M., Heine, C., Suazo, C., et al., 2012. Mutación homocigota en la línea germinal del gen MUTYH en una paciente chilena con poliposis adenomatosa familiar. Revista médica de Chile. scielocl 140, 1457–1463.

Ashton-Prolla, P., Seuanez, H.N., 2016. The Brazilian Hereditary Cancer Network: historical aspects and challenges for clinical cancer genetics in the public health care system in Brazil. Genet. Mol. Biol. 39 (2), 163–165.

Ashton-Prolla, P., Vargas, F.R., 2014. Prevalence and impact of founder mutations in hereditary breast cancer in Latin America. Genet. Mol. Biol. 37 (1 Suppl. 1), 234–240.

Berg, A.O., Armstrong, K., Botkin, J., Calonge, N., Haddow, J., Hayes, M., et al., 2009. Recommendations from the EGAPP Working Group: genetic testing strategies in newly diagnosed individuals with colorectal cancer aimed at reducing morbidity and mortality from Lynch syndrome in relatives. Genet. Med. 11 (1), 35–41.

Bertario, L., Russo, A., Sala, P., Varesco, L., Giarola, M., Mondini, P., et al., 2003. Multiple approach to the exploration of genotype-phenotype correlations in familial adenomatous polyposis. J. Clin. Oncol. 21 (9), 1698–1707.

Bisgaard, M.L., Bülow, S., 2006. Familial adenomatous polyposis (FAP): genotype correlation to FAP phenotype with osteomas and sebaceous cysts. Am. J. Med. Genet. 140 (A(3)), 200–204.

Boland, C.R., Troncale, F.J., 1984. Familial colonic cancer without antecedent polyposis. Ann. Int. Med. 100 (5), 700–701.

Bonilla, C., Bertoni, B., Hidalgo, P.C., Artagaveytia, N., Ackermann, E., Barreto, I., et al., 2015. Breast cancer risk and genetic ancestry: a case-control study in Uruguay. BMC Womens Health 15 (1), 11.

Cajal, A.R., Pinero, T.A., Verzura, A., Santino, J.P., Solano, A.R., Kalfayan, P.G., et al., 2016. Founder mutation in Lynch syndrome. Medicina (B Aires) 76 (3), 180–182.

Carneiro da Silva, F., Ferreira, J.R., de, O., Torrezan, G.T., Figueiredo, M.C.P., Santos ÉMM, Nakagawa, W.T., et al., 2015. Clinical and molecular characterization of Brazilian patients suspected to have Lynch syndrome. PLoS One 10 (10), e0139753.

Carraro, D.M., Koike Folgueira, M.A.A., Garcia Lisboa, B.C., Ribeiro Olivieri, E.H., Vitorino Krepischi, A.C., de Carvalho, A.F., et al., 2013. Comprehensive analysis of *BRCA1BRCA2* and TP53 germline mutation and tumor characterization: a portrait of early-onset breast cancer in Brazil. PLoS One 8 (3).

Caspari, R., Friedl, W., Mandl, M., Möslem, G., Kadmon, M., Knapp, M., et al., 1994. Familial adenomatous polyposis: mutation at codon 1309 and early onset of colon cancer. Lancet 343 (8898), 629–632.

Caspari, R., Olschwang, S., Friedl, W., Mandl, M., Boisson, C., Böker, T., et al., 1995. Familial adenomatous polyposis: desmoid tumours and lack of ophthalmic lesions (chrpe) associated with APC mutations beyond codon 1444. Hum. Mol. Genet. 4 (3), 337–340.

Castro-Mujica Mdel, C., Sullcahuaman-Allende, Y., Barreda-Bolanos, F., Taxa-Rojas, L., 2014. Inherited colorectal cancer predisposition syndromes identified in the Instituto Nacional de Enfermedades Neoplasicas (INEN), Lima, Peru. Rev. Gastroenterol. Peru 34 (2), 107–114.

Castro-Mujica, M.D.C., Barletta-Carrillo, C., Acosta-Aliaga, M., Montenegro-Garreaud, X., 2016. Lynch syndrome Muir Torre variant: 2 cases. Rev. Gastroenterol. Peru 36 (1), 81–85.

Cheadle, J.P., Sampson, J.R., 2003. Exposing the MYtH about base excision repair and human inherited disease. Hum. Mol. Genet. 12, Spec No (2):R159-65.

Chen, S., Parmigiani, G., 2007. Meta-analysis of *BRCA1* and *BRCA2* penetrance. J. Clin. Oncol. 25 (11), 1329–1333.

Chialina, S.G., Fornes, C., Landi, C., de la Vega Elena, C.D., Nicolorich, M.V., Dourisboure, R.J., et al., 2006. Microsatellite instability analysis in hereditary non-polyposis colon cancer using the Bethesda consensus panel of microsatellite markers in the absence of proband normal tissue. BMC Med. Genet. 7, 5.

Clarizia, A.D., Bastos-Rodrigues, L., Pena, H.B., Anacleto, C., Rossi, B., Soares, F.A., et al., 2006. Relationship of the methylenetetrahydrofolate reductase C677T polymorphism with microsatellite instability and promoter hypermethylation in sporadic colorectal cancer. Genet. Mol. Res. 5 (2), 315–322.

Cruz-Bustillo, D., Villasana, L., Llorente, F., Casadesús, D., García, E., Syrris, P., et al., 2002. Preliminary results of the molecular diagnosis of familial adenomatous polyposis in Cuban families. Int. J. Colorectal Dis. 17 (5), 344–347.

Cruz-Correa, M., Diaz-Algorri, Y., Mendez, V., Vazquez, P.J., Lozada, M.E., Freyre, K., et al., 2013. Clinical characterization and mutation spectrum in Hispanic families with adenomatous polyposis syndromes. Fam. Cancer 12 (3), 555–562.

Cruz-Correa, M., Diaz-Algorri, Y., Pérez-Mayoral, J., Suleiman-Suleiman, W., del Mar Gonzalez-Pons, M., Bertrán, C., et al., 2015. Clinical characterization and mutation spectrum in Caribbean Hispanic families with Lynch syndrome. Fam. Cancer 14 (3), 415–425.

Da Silva, F.C., Wernhoff, P., Dominguez-Barrera, C., Dominguez-Valentin, M., 2016. Update on hereditary colorectal cancer. Anticancer Res. 36, 4399–4406.

de Freitas, I.N., de Campos, F.G.C.M., Alves, V.A.F., Cavalcante, J.M., Carraro, D., Coudry, R., de, A., et al., 2015. Proficiency of DNA repair genes and microsatellite instability in operated colorectal cancer patients with clinical suspicion of lynch syndrome. J. Gastrointest Oncol. 6 (6), 628–637.

De Jesus-Monge, W.E., Gonzalez-Keelan, C., Zhao, R., Hamilton, S.R., Rodriguez-Bigas, M., Cruz-Correa, M., 2010. Mismatch repair protein expression and colorectal cancer in Hispanics from Puerto Rico. Fam. Cancer 9 (2), 155–166.

De La Fuente, M.K., Alvarez, K.P., Letelier, A.J., Bellolio, F., Acuña, M.L., León, F.S., et al., 2007. Mutational screening of the APC gene in Chilean families with familial adenomatous polyposis: nine novel truncating mutations. Dis. Colon Rectum 50 (12), 2142–2148.

De Queiroz Rossanese, L.B., De Lima Marson, F.A., Ribeiro, J.D., Coy, C.S.R., Bertuzzo, C.S., 2013. APC germline mutations in families with familial adenomatous polyposis. Oncol. Rep. 30 (5), 2081–2088.

De Rosa, M., Dourisboure, R.J., Morelli, G., Graziano, A., Gutiérrez, A., Thibodeau, S., et al., 2004. First genotype characterization of Argentinean FAP patients: identification of 14 novel APC mutations. Hum. Mutat. 23 (5), 523–524.

Delgado, L., Fernández, G., Grotiuz, G., Cataldi, S., González, A., Lluveras, N., et al., 2011. *BRCA1* and *BRCA2* germline mutations in Uruguayan breast and breast-ovarian cancer families. Identification of novel mutations and unclassified variants. Breast Cancer Res. Treat. 128 (1), 211–218.

Dominguez-Valentin, M., Nilbert, M., Wernhoff, P., López-Köstner, F., Vaccaro, C., Sarroca, C., et al., 2013. Mutation spectrum in South American Lynch syndrome families. Hered Cancer Clin. Pract. 11 (1), 18.

Dominguez-Valentin, M., Therkildsen, C., Da Silva, S., Nilbert, M., 2015. Familial colorectal cancer type X: genetic profiles and phenotypic features. Mod. Pathol. 28 (1), 30–36.

Dominguez-Valentin, M., Wernhoff, P., Cajal, A.R., Kalfayan, P.G., Piñero, T.A., Gonzalez, M.L., et al., 2016. MLH1 Ile219Val polymorphism in Argentinean families with suspected Lynch syndrome. Front. Oncol. 6, 189.

Egoavil, C.M., Montenegro, P., Soto, J.L., Casanova, L., Sanchez-Lihon, J., Castillejo, M.I., et al., 2011. Clinically important molecular features of Peruvian colorectal tumours: high prevalence of DNA mismatch repair deficiency and low incidence of KRAS mutations. Pathology 43 (3), 228–233.

Ewald, I.P., Cossio, S.L., Palmero, E.I., Pinheiro, M., Nascimento, I.L., Machado, T.M., et al., 2016. *BRCA1* and *BRCA2* rearrangements in Brazilian individuals with Hereditary Breast and Ovarian Cancer Syndrome. Genet. Mol. Biol. 39 (2), 223–231.

Eyheramendy, S., Martinez, F.I., Manevy, F., Vial, C., Repetto, G.M., 2015. Genetic structure characterization of Chileans reflects historical immigration patterns. Nat. Commun., 6.

Fearnhead, N.S., Britton, M.P., Bodmer, W.F., Hospital, J.R., Ox, O., 2001. The ABC of APC. Hum. Mol. Genet. 10 (7), 721–733.

Fejerman, L., John, E.M., Huntsman, S., Beckman, K., Choudhry, S., Perez-Stable, E., et al., 2008. Genetic ancestry and risk of breast cancer among U.S. Latinas. Cancer Res. 68 (23), 9723–9728.

Fejerman, L., Romieu, I., John, E.M., Lazcano-Ponce, E., Huntsman, S., Beckman, K.B., et al., 2010. European ancestry is positively associated with breast cancer risk in Mexican women. Cancer Epidemiol. Biomarkers Prev. 19 (4), 1074–1082.

Fejerman, L., Ahmadiyeh, N., Hu, D., Huntsman, S., Beckman, K.B., Caswell, J.L., et al., 2014. Genome-wide association study of breast cancer in Latinas identifies novel protective variants on 6q25. Nat. Commun. 5, 5260.

Felix, G.E., Abe-Sandes, C., Machado-Lopes, T.M., Bomfim, T.F., Guindalini, R.S.C., Santos, V.C.S., et al., 2014. Germline mutations in *BRCA1*, *BRCA2*, CHEK2 and TP53 in patients at high-risk for HBOC: characterizing a Northeast Brazilian Population. Hum. Genome Var. 1, 14012.

Fernandes, G.C., Michelli, R.A.D., Galvão, H.C.R., Paula, A.E., Pereira, R., Andrade, C.E., et al., 2016. Prevalence of *BRCA1/BRCA2* mutations in a Brazilian population sample at-risk for hereditary breast cancer and characterization of its genetic ancestry. Oncotarget 7 (49), 80465–80481.

Friedl, W., Meuschel, S., Caspari, R., Lamberti, C., Krieger, S., Sengteller, M., et al., 1996. Attenuated familial adenomatous polyposis due to a mutation in the 3′ part of the APC gene A clue for understanding the function of the APC protein. Hum. Genet. 97 (5), 579–584.

Galiatsatos, P., Foulkes, W.D., 2006. Familial adenomatous polyposis. Am. J. Gastroenterol. 101 (2), 385–398.

Gallardo, M., Silva, A., Rubio, L., Alvarez, C., Torrealba, C., Salinas, M., et al., 2006. Incidence of *BRCA1* and *BRCA2* mutations in 54 Chilean families with breast/ovarian cancer, genotype-phenotype correlations. Breast Cancer Res. Treat. 95 (1), 81–87.

Garcia, G.H., Riechelmann, R.P., Hoff, P.M., 2015. Adherence to colonoscopy recommendations for first-degree relatives of young patients diagnosed with colorectal cancer. Clinics 70 (10), 696–699.

Germini, D.E., Mader, A.M.A.A., Gomes, L.G.L., Teodoro, T.R., Franco, M.I.F., Waisberg, J., 2016. Detection of DNA repair protein in colorectal cancer of patients up to 50 years old can increase the identification of Lynch syndrome? Tumor Biol. 37 (2), 2757–2764.

Giraldo, A., Gómez, A., Salguero, G., García, H., Aristizábal, F., Gutiérrez, Ó., et al., 2005. MLH1 and MSH2 mutations in Colombian families with hereditary nonpolyposis colorectal cancer (Lynch syndrome) – Description of four novel mutations. Fam. Cancer 4 (4), 285–290.

Gonzalez-Hormazabal, P., Gutierrez-Enriquez, S., Gaete, D., Reyes, J.M., Peralta, O., Waugh, E., et al., 2011. Spectrum of *BRCA1/2* point mutations and genomic rearrangements in high-risk breast/ovarian cancer Chilean families. Breast Cancer Res. Treat. 126 (3), 705–716.

Groden, J., Thliveris a, Samowitz, W., Carlson, M., Gelbert, L., Albertsen, H., et al., 1991. Identification and characterization of the familial adenomatous polyposis coli gene. Cell 66 (3), 589–600.

Gutiérrez Espeleta, G., Llacuachaqui, M., García-Jiménez, L., Aguilar Herrera, M., Loáiciga Vega, K., Ortiz, A., et al., 2012. *BRCA1* and *BRCA2* mutations among familial breast cancer patients from Costa Rica. Clin. Genet. 82 (5), 484–488.

Hall, J.M., Lee, M.K., Newman, B., Morrow, J.E., Anderson, L.A., Huey, L.A., et al., 1990. Linkage of early-onset familial breast cancer to chromosome 17q21. Science (80) 250, 1684–1689.

Hamel, N., Feng, B.-J., Foretova, L., Stoppa-Lyonnet, D., Narod, S.A., Imyanitov, E., et al., 2011. On the origin and diffusion of BRCA1 c 5266dupC (5382insC) in European populations. Eur. J. Hum. Genet. 19 (3), 300–306.

Hampel, H., Frankel, W.L., Martin, E., Arnold, M., Khanduja, K., Kuebler, P., et al., 2005. Screening for the Lynch syndrome (hereditary nonpolyposis colorectal cancer). N. Engl. J. Med. 352 (18), 1851–1860.

Hampel, H., Frankel, W., Panescu, J., Lockman, J., Sotamaa, K., Fix, D., et al., 2006. Screening for Lynch syndrome (hereditary nonpolyposis colorectal cancer) among endometrial cancer patients. Cancer Res. 66 (15), 7810–7817.

He, T., Sparks, A.B., Rago, C., Hermeking, H., Zawel, L., Costa, L.T., et al., 1998. Identification of c- MYC as a target of the APC pathway. Science (80) 281, 1509–1512.

Hernández, J.E.L., Llacuachaqui, M., Palacio, G.V., Figueroa, J.D., Madrid, J., Lema, M., et al., 2014. Prevalence of BRCA1 and BRCA2 mutations in unselected breast cancer patients from medellín Colombia. Hered. Cancer Clin. Pract. 12 (1), 11.

Howlader, N., Noone, A.M., Krapcho, M., et al., 2016. SEER cancer statistics review, 1975–2013. In: National Cancer Institute Bethesda, MD.

Jara, L., Ampuero, S., Santibáñez, E., Seccia, L., Rodríguez, J., Bustamante, M., et al., 2006. BRCA1 and BRCA2 mutations in a South American population. Cancer Genet. Cytogenet. 166 (1), 36–45.

Kinzler, K., Nilbert, M., Su, L., Vogelstein, B., Bryan, T., Levy, D., et al., 1991. Identification of FAP locus genes from chromosome 5q21. Science (80) 253 (5020), 661–665.

Knudsen, A.L., Bisgaard, M.L., Bülow, S., 2003. Attenuated familial adenomatous polyposis (AFAP). A review of the literature. Familial Cancer 2, 43–55.

Kobayashi, H., Ohno, S., Sasaki, Y., Matsuura, M., 2013. Hereditary breast and ovarian cancer susceptibility genes (Review). Oncol. Rep. 30, 1019–1029.

Koehler-Santos, P., Izetti, P., Abud, J., Pitroski, C.E., Cossio, S.L., Camey, S.A., et al., 2011. Identification of patients at-risk for lynch syndrome in a hospital-based colorectal surgery clinic. World J. Gastroenterol. 17 (6), 766–773.

Lara, K., Consigliere, N., Pérez, J., Porco, A., 2012. BRCA1 and BRCA2 mutations in breast cancer patients from Venezuela. Biol. Res. 45 (2), 117–130.

Leite, S.M.O., Gomes, K.B., Pardini, V.C., Ferreira, A.C.S., Oliveira, V.C., Cruz, G.M.G., 2010. Assessment of microsatellite instability in colorectal cancer patients from Brazil. Mol. Biol. Rep. 37 (1), 375–380.

Leppert, M., Dobbs, M., Scambler, P., O'Connell, P., Nakamura, Y., Stauffer, D., et al., 1987. The gene for familial polyposis coli maps to the long arm of chromosome 5. Science (80) 238 (4832), 1411–1413.

Marqués-Lespier, J.M., Diaz-Algorri, Y., Gonzalez-Pons, M., Cruz-Correa, M., 2014. Report of a novel mutation in MLH1 gene in a Hispanic family from Puerto Rico fulfilling classic Amsterdam criteria for lynch syndrome. Gastroenterol. Res. Pract. 2014.

Merg, A., Lynch, H.T., Lynch, J.F., Howe, J.R., 2005. Hereditary colon cancer--part I. Curr. Prob. Surg. 42 (4), 195–256.

Miki, Y., Swensen, J., Shattuck-Eidens, D., Futreal, P.A., Harshman, K., Tavtigian, S., et al., 1994. A strong candidate for the breast and ovarian cancer susceptibility gene BRCA1. Science 266 (5182), 66–71.

Milne, R.L., Osorio, A., Cajal, T.R.Y., Vega, A., Llort, G., De La Hoya, M., et al., 2008. The average cumulative risks of breast and ovarian cancer for carriers of mutations in BRCA1 and BRCA2 attending genetic counseling units in Spain. Clin. Cancer Res. 14 (9), 2861–2869.

Møller, P., Seppälä, T., Bernstein, I., Holinski-Feder, E., Sala, P., Evans, D.G., et al., 2015. Cancer incidence and survival in Lynch syndrome patients receiving colonoscopic and gynaecological surveillance: first report from the prospective Lynch syndrome database. Gut.

Møller, P., Seppälä, T., Bernstein, I., Holinski-Feder, E., Sala, P., Evans, D.G., et al., 2016. Incidence of and survival after subsequent cancers in carriers of pathogenic MMR variants with previous cancer: a report from the prospective Lynch syndrome database. Gut, gutjnl-2016-311403.

Montenegro, Y., Ramirez-Castro, J.L., Isaza, L.F., Bedoya, G., Muneton-Pena, C.M., 2006. Microsatellite instability among patients with colorectal cancer. Rev. Med. Chil. 134 (10), 1221–1229.

Moreno-Ortiz, J.M., Ayala-Madrigal, M., de la, L., Corona-Rivera, J.R., Centeno-Flores, M., Maciel-Gutiérrez, V., Franco-Topete, R.A., et al., 2016. Novel Mutations in *MLH1* and *MSH2* Genes in Mexican Patients with Lynch Syndrome. Gastroenterol. Res. Pract. 2016, 1–6.

Nagase, H., Miyoshi, Y., Horii, A., Aoki, T., Ogawa, M., Utsunomiya, J., et al., 1992. Correlation between the location of germ-line mutations in the APC gene and the number of colorectal polyps in familial adenomatous polyposis patients. Cancer Res. 52 (14), 4055–4057.

Nakamura, Y., Lathrop, M., Leppert, M., Dobbs, M., Wasmuth, J., Wolff, E., et al., 1988. Localization of the genetic defect in familial adenomatous polyposis within a small region of chromosome 5. Am. J. Hum. Genet. 43 (5), 638–644.

Nique Carbajal, C., Sanchez Renteria, F., Lettiero, B., Wernhoff, P., Dominguez-Valentin, M., 2014. Molecular characterization of hereditary colorectal cancer in Peru. Rev. Gastroenterol. Peru 34 (4), 299–303.

Ohtsubo, T., 2000. Identification of human MutY homolog (hMYH) as a repair enzyme for 2-hydroxyadenine in DNA and detection of multiple forms of hMYH located in nuclei and mitochondria. Nucleic Acids Res. 28 (6), 1355–1364.

Oka, S., Nakabeppu, Y., 2011. DNA glycosylase encoded by MUTYH functions as a molecular switch for programmed cell death under oxidative stress to suppress tumorigenesis. Cancer Sci. 102, 677–682.

Ortiz, C., Dongo-Pflucker, K., Martin-Cruz, L., Barletta Carrillo, C., Mora-Alferez, P., Arias, A., 2016. Microsatellite instability in patients with diagnostic of colorectal cancer. Rev. Gastroenterol. Peru 36 (1), 15–22.

Ossa, C.A., Torres, D., 2016. Founder and recurrent mutations in *BRCA1* and *BRCA2* genes in latin american countries: state of the art and literature review. Oncologist 21 (7), 832–839.

Ramírez-Ramírez, M.A., Sobrino-Cossío, S., De La Mora-Levy, J.G., Hernández-Guerrero, A., Macedo-Reyes, V.D.J., Maldonado-Martínez, H.A., et al., 2012. Loss of expression of DNA mismatch repair proteins in aberrant crypt foci identified in vivo by magnifying colonoscopy in subjects with hereditary nonpolyposic and sporadic colon rectal cancer. J. Gastrointest Cancer 43 (2), 209–214.

Rasuck, C.G., Leite, S.M.O., Komatsuzaki, F., Ferreira, A.C.S., Oliveira, V.C., Gomes, K.B., 2012. Association between methylation in mismatch repair genes, V600E BRAF mutation and microsatellite instability in colorectal cancer patients. Mol. Biol. Rep. 39 (3), 2553–2560.

Ricker, C., Klipfel, N., Ault, G., Roman, L., Spicer, D., Lenz, H.-J., 2010a. Characteristics of Lynch syndrome in 13 Hispanic Families. Hered. Cancer Clin. Pract. 8 (Suppl. 1), 19–119.

Ricker, C., Ault, G., El-Khoureiy, A., Iqbal, S., Spicer, D., Lenz, H.-J., 2010b. Familial adenomatous polyposis (FAP) in 9 Hispanic women. Hered. Cancer Clin. Pract. 8(Suppl. 1), P18–P118.

Rocco, P., Morales, C., Moraga, M., Miquel, J.F., Nervi, F., Llop, E., et al., 2002. Genetic composition of the Chilean population. Analysis of mitochondrial DNA polymorphism. Rev. Med. Chil. 130 (2), 125–131.

Rodriguez, R.C., Esperon, A.A., Ropero, R., Rubio, M.C., Rodriguez, R., Ortiz, R.M., et al., 2008. Prevalence of *BRCA1* and *BRCA2* mutations in breast cancer patients from Cuba. Fam. Cancer 7 (3), 275–279.

Rodriguez-Bigas, M.A., Boland, C.R., Hamilton, S.R., Henson, D.E., Jass, J.R., Khan, P.M., et al., 1997. A National Cancer Institute Workshop on Hereditary Nonpolyposis Colorectal Cancer Syndrome: meeting highlights and Bethesda guidelines. J. Natl. Cancer Inst. United States 89, 1758–1762.

Roque, M., Pusiol, E., Giribet, G., Perinetti, H., Mayorga, L.S., 2000. Diagnosis by directed mutagenesis of a mutation at the hMSH2 gene associated with hereditary nonpolyposis colorectal cancer. Medicina (B Aires) 60 (2), 188–194.

Rossi, B.M., Lopes, A., Oliveira Ferreira, F., Nakagawa, W.T., Napoli Ferreira, C.C., Casali Da Rocha, J.C., et al., 2002. hMLH1 and hMSH2 gene mutation in Brazilian families with suspected hereditary nonpolyposis colorectal cancer. Ann. Surg. Oncol. 9 (6), 555–561.

Rossi, B.M., Sarroca, C., Vaccaro, C., Lopez, F., Ashton-Prolla, P., Ferreira, F., de, O., et al., 2016. The development of the study of hereditary cancer in South America. Genet. Mol. Biol. 39 (2), 166–167.

Rossi, B.M., Palmero, E.I., López-Kostner, F., Sarroca, C., Vaccaro, C.A., Spirandelli, F., et al., 2017. A survey of the clinicopathological and molecular characteristics of patients with suspected Lynch syndrome in Latin America. BMC Cancer 17, 623.

Rubenstein, J.H., Enns, R., Heidelbaugh, J., Barkun, A., Adams, M.A., Dorn, S.D., et al., 2015. American gastroenterological association institute guideline on the diagnosis and management of Lynch syndrome. Gastroenterology 149 (3), 777–782.

Ruiz-Linares, A., Adhikari, K., Acuña-Alonzo, V., Quinto-Sanchez, M., Jaramillo, C., Arias, W., et al., 2014. Admixture in Latin America: geographic structure, phenotypic diversity and self-perception of ancestry based on 7,342 individuals. PLoS Genet. 10 (9).

Sampson, J.R., Dolwani, S., Jones, S., Eccles, D., Ellis, A., Evans, D.G., et al., 2003. Autosomal recessive colorectal adenomatous polyposis due to inherited mutations of MYH. Lancet 362 (9377), 39–41.

Sanchez, A., Faundez, P., Carvallo, P., 2011. Genomic rearrangements of the *BRCA1* gene in Chilean breast cancer families: an MLPA analysis. Breast Cancer Res. Treat. 128 (3), 845–853.

Santos, E.M.M., Vatentin, M.D., Carneiro, F., Oliveira, L.P., Ferreira, F.O., Nakagawa, W.T., et al., 2012. Predictive models for mutations in mismatch repair genes: implication for genetic counseling in developing countries. BMC Cancer 12 (1), 64.

Sarroca, C., Valle a Della, Fresco, R., Renkonen, E., Peltömaki, P., Lynch, H., 2005. Frequency of hereditary non-polyposis colorectal cancer among Uruguayan patients with colorectal cancer. Clin. Genet. 68 (1), 80–87.

Sjursen, W., Haukanes, B.I., Grindedal, E.M., Aarset, H., Stormorken, A., Engebretsen, L.F., et al., 2010. Current clinical criteria for Lynch syndrome are not sensitive enough to identify MSH6 mutation carriers. J. Med. Genet. 47 (9), 579–585.

Sjursen, W., McPhillips, M., Scott, R.J., Talseth-Palmer, B.A., 2016. Lynch syndrome mutation spectrum in New South Wales, Australia, including 55 novel mutations. Mol. Genet. Genomic Med. 4 (2), 223–231.

Solano, A.R., Aceto, G.M., Delettieres, D., Veschi, S., Neuman, M.I., Alonso, E., et al., 2012. *BRCA1* And *BRCA2* analysis of Argentinean breast/ovarian cancer patients selected for age and family history highlights a role for novel mutations of putative south-American origin. Springerplus 1, 20.

Solano, A.R., Cardoso, F.C., Romano, V., Perazzo, F., Bas, C., Recondo, G., et al., 2016. Spectrum of *BRCA1/2* variants in 940 patients from Argentina including novel, deleterious and recurrent germline mutations: impact on healthcare and clinical practice. Oncotarget.

Spirio, L., Olschwang, S., Groden, J., Robertson, M., Samowitz, W., Joslyn, G., et al., 1993. Alleles of the APC gene: an attenuated form of familial polyposis. Cell 75 (5), 951–957.

Strate, L.L., Syngal, S., 2005. Hereditary colorectal cancer syndromes. Cancer Causes Control 16, 201–213.

Theodoratou, E., Campbell, H., Tenesa, A., Houlston, R., Webb, E., Lubbe, S., et al., 2010. A large-scale meta-analysis to refine colorectal cancer risk estimates associated with MUTYH variants. Br. J. Cancer 103 (12), 1875–1884.

Thompson, B.A., Spurdle, A.B., Plazzer, J.-P., Greenblatt, M.S., Akagi, K., Al-Mulla, F., et al., 2014. Application of a 5-tiered scheme for standardized classification of 2,360 unique mismatch repair gene variants in the InSiGHT locus-specific database. Nat. Genet. 46 (2), 107–115.

Tiwari, A.K., Roy, H.K., Lynch, H.T., 2016. Lynch syndrome in the 21st century: clinical perspectives. QJM 109 (3), 151–158.

Torres, D., Rashid, M.U., Gil, F., Umana, A., Ramelli, G., Robledo, J.F., et al., 2007. High proportion of BRCA1/2 founder mutations in Hispanic breast/ovarian cancer families from Colombia. Breast Cancer Res. Treat. 103 (2), 225–232.

Torres-Mejia, G., Royer, R., Llacuachaqui, M., Akbari, M.R., Giuliano, a.R., Martinez-Matsushita, L., et al., 2015. Recurrent BRCA1 and BRCA2 mutations in Mexican women with breast cancer. Cancer Epidemiol. Biomarkers Prev. 24 (3), 498–505.

Torrezan, G.T., da Silva, F.C.C., Santos, E.M.M., Krepischi, A.C.V., Achatz, M.I.W., Aguiar, S., et al., 2013. Mutational spectrum of the APC and MUTYH genes and genotype-phenotype correlations in Brazilian FAP, AFAP, and MAP patients. Orphanet J. Rare Dis. 8, 54.

Umar, A., Boland, C.R., Terdiman, J.P., Syngal, S., de la Chapelle, A., Rüschoff, J., et al., 2004. Revised Bethesda Guidelines for hereditary nonpolyposis colorectal cancer (Lynch syndrome) and microsatellite instability. J. Natl. Cancer Inst. 96 (4), 261–268.

Vaca-Paniagua, F., Alvarez-Gomez, R.M., Fragoso-Ontiveros, V., Vidal-Millan, S., Herrera, L.A., Cantú, D., et al., 2012. Full-Exon pyrosequencing screening of BRCA germline mutations in Mexican women with inherited breast and ovarian cancer. PLoS One 7 (5).

Vaccaro, C.A., Bonadeo, F., Roverano, A.V., Peltomaki, P., Bala, S., Renkonen, E., et al., 2007a. Hereditary nonpolyposis colorectal cancer (Lynch syndrome) in Argentina: report from a referral hospital register. Dis. Colon Rectum 50 (10), 1604–1611.

Vaccaro, C.A., Carrozzo, J.E., Mocetti, E., Berho, M., Valdemoros, P., Mullen, E., et al., 2007b. Immunohistochemical expression and microsatellite instability in Lynch syndrome. Medicina (B Aires) 67 (3), 274–278.

Vaccaro, C.A., Sarroca, C., Rossi, B., Lopez-Kostner, F., Dominguez, M., Calo, N.C., et al., 2016. Lynch syndrome in South America: past, present and future. Fam. Cancer 15 (3), 437–445.

Valentin, M.D., Da Silva, F.C., Santos, E.M.M., Dos, Lisboa, B.G., De Oliveira, L.P., De Oliveira Ferreira, F., et al., 2011. Characterization of germline mutations of MLH1 and MSH2 in unrelated south American suspected Lynch syndrome individuals. Fam. Cancer 10 (4), 641–647.

Valentin, M.D., Da Silva, F.C., Santos, E.M.M., Da Silva, S.D., De Oliveira Ferreira, F., Aguiar, S., et al., 2012. Evaluation of MLH1 I219V polymorphism in unrelated South American individuals suspected of having Lynch syndrome. Anticancer Res. 32 (10), 4347–4352.

Vasen, H.F., Muller, H., 1991. DNA studies in families with hereditary forms of cancer. Ned Tijdschr Geneeskd. 135 (36), 1620–1623.

Vasen, H.F.A., Watson, P., Mecklin, J.P., Lynch, H.T., 1999. New clinical criteria for hereditary non-polyposis colorectal cancer (HNPCCLynch syndrome) proposed by the International Collaborative Group on HNPCC. Gastroenterology, 1453–1456.

Viana, D.V., Góes, J.R.N., Coy, C.S.R., De Lourdes Setsuko Ayrizono, M., Lima, C.S.P., Lopes-Cendes, I., 2008. Family history of cancer in Brazil: Is it being used? Fam. Cancer 7 (3), 229–232.

Villarreal-Garza, C., Alvarez-Gómez, R.M., Pérez-Plasencia, C., Herrera, L.A., Herzog, J., Castillo, D., et al., 2015a. Significant clinical impact of recurrent *BRCA1* and *BRCA2* mutations in Mexico. Cancer 121 (3), 372–378.

Villarreal-Garza, C., Weitzel, J.N., Llacuachaqui, M., Sifuentes, E., Magallanes-Hoyos, M.C., Gallardo, L., et al., 2015b. The prevalence of *BRCA1* and *BRCA2* mutations among young Mexican women with triple-negative breast cancer. Breast Cancer Res. Treat. 150 (2), 389–394.

Weitzel, J.N., Lagos, V.I., Herzog, J.S., Judkins, T., Hendrickson, B., Ho, J.S., et al., 2007. Evidence for common ancestral origin of a recurring *BRCA1* genomic rearrangement identified in high-risk hispanic families. Cancer Epidemiol. Biomarkers Prev. 16 (8), 1615–1620.

Weitzel, J.N., Clague, J., Martir-Negron, A., Ogaz, R., Herzog, J., Ricker, C., et al., 2013. Prevalence and type of BRCA mutations in Hispanics undergoing genetic cancer risk assessment in the southwestern United States: a report from the Clinical Cancer Genetics Community Research Network. J. Clin. Oncol. 31 (2), 210–216.

Wielandt, A.M., Zárate, A.J., Hurtado, C., Orellana, P., Alvarez, K., Pinto, E., et al., 2012. Lynch syndrome: selection of families by microsatellite instability and immunohistochemistry. Rev. Méd. Chil. 140 (9), 1132–1139.

Wooster, R., Bignell, G., Lancaster, J., Swift, S., Seal, S., Mangion, J., et al., 1995. Identification of the breast cancer susceptibility gene *BRCA2*. Nature 378 (6559), 789–792.

# CHAPTER 6

# Implementing Genomics in the Care of Neuropsychiatric Patients in Latin America

**Diego A. Forero\*, Hermes Urriago\*, Sandra Lopez-Leon\*\*, Yeimy González-Giraldo†, Debora M. de Miranda‡, Camilo A. Espinosa Jovel§**

*\*Universidad Antonio Nariño, Bogotá, Colombia; \*\*Novartis Pharmaceuticals Corporation, East Hanover, NJ, United States; †Pontificia Universidad Javeriana, Bogotá, Colombia; ‡Universidade Federal de Minas Gerais, Belo Horizonte, Brazil; §Hospital de Kennedy, Bogotá, Colombia*

## INTRODUCTION

Neuropsychiatric disorders (NPDs) lead to a high disease burden in Latin American countries (Kohn et al., 2005; Whiteford et al., 2015). An understanding of the multiple risk and causal factors for NPDs is of particular importance, considering that some of these entities have a significant heritability (Zoghbi and Warren, 2010). In this chapter, we review the previous and current advances in research into neuropsychiatric genetics (NPG) in Latin America (Cordeiro et al., 2009b; Forero et al., 2014).

### Epidemiological Impact of Neurological Diseases and Psychiatric Disorders in Latin America

NPDs have an important impact on the total mortality, morbidity, and disability in all stages of life; they represent 22% of the total burden of disease in Latin America. Previous work from the World Health Organization (for example, the Global Burden of Disease Study) has shown that psychiatric disorders (PDs) and neurological diseases (NDs) represent a huge challenge for global public health and biomedical research, in terms of years lived with disability (YLDs) and disability-adjusted life years (DALYs) (Demyttenaere et al., 2004; Disease, Injury, & Prevalence, 2016; Vigo et al., 2016; Whiteford et al., 2015). In relationship to DALYs, major depressive disorder (MDD), anxiety disorders, migraine, alcohol dependence, epilepsy, schizophrenia (SZ), bipolar disorder (BP), and Alzheimer's disease (AD) and other dementias have the highest impact (Whiteford et al., 2015).

Genomic Medicine in Emerging Economies. http://dx.doi.org/10.1016/B978-0-12-811531-2.00006-9

Mental, neurological, and substance use disorders accounted for 258 million DALYs in 2010, which was equivalent to 10.4% of total all-cause DALYs (Murray et al., 2012; Whiteford et al., 2015). These disorders as a group ranked as the third leading cause of DALYs, after cardiovascular and circulatory diseases, diarrhea, lower respiratory infections, meningitis, and other common infectious diseases (Murray et al., 2012; Whiteford et al., 2015). NPDs also accounted for 28.5% of global YLDs, making them the leading cause of YLDs (Murray et al., 2012; Whiteford et al., 2015). The distribution of DALYs among NPDs is heterogeneous and depends mainly on the economic profile of the population studied; for example, epilepsy ranks as the 36th leading cause of DALYs globally; however, in some low to middle income countries of Latin America and western sub-Saharan Africa, NPDs rank as the 21th and 14th leading cause of DALYs, respectively (Murray et al., 2012). Epilepsy also represents the second most disabling neurological disorder (as measured in YLDs), only surpassed by migraine, and in some Latin American countries it ranks as the ninth leading cause of YLDs (Vos et al., 2012).

Several studies have been carried out in Latin America to estimate the prevalence in the general population of selected NPDs (Table 6.1) (Burneo et al., 2005; Cristiano et al., 2013; Guerra et al., 2009; Jimenez-Castro et al., 2011; Kisely et al., 2017; Kohn et al., 2005; Kolar et al., 2016; Nitrini et al., 2009; Prince et al., 2013; Pringsheim et al., 2014), highlighting the national and regional importance of several NPDs of high epidemiological impact.

The population of Latin America and the Caribbean is composed of approximately 620 million people living in 26 countries, speaking two major languages, and sharing ancestral roots in America, Europe, and Africa (Adhikari et al., 2016) (Fig. 6.1). The region is considered one of the most diverse in the world, with an extensive variability in ancestries, ethnic groups, and races (Li et al., 2008). The population of Latin America is the result of the mating of Native Americans with individuals from early migrations from Africa and Europe and later waves of European and Asian migration (Adhikari et al., 2016). Recent analyses from mestizo populations from seven countries in Latin America showed evidence that the genetic origins of these populations are composed mainly by European men and Native American women (Wang et al., 2008). In this study, the Native American ancestry ranged from ∼20% to ∼70% in Rio Grande do Sul and Salta, respectively (Wang et al., 2008). The African ancestry is also variable and mostly low (<5%) and possibly related to the intensity of slavery in each country (Wang et al., 2008). Considering that Brazil had a long and intense history of slavery, the frequency of African ancestry is much higher than the mean frequency found in other countries in Latin America. The pooled ancestry frequency from African ancestry in Brazil was 21% (Moura et al., 2015). However, unexpectedly, in a study involving 934

**Table 6.1** An Overview of Prevalence of Psychiatric Disorders and Neurological Diseases in Latin American Countries

| Outcome | Region | Study Period | Prevalence Description | Prevalence % | References |
|---|---|---|---|---|---|
| **Psychiatric disorders** | | | | | |
| Alcohol dependence and harmful alcohol use | Latin America and Caribbean | 1980–2004 | Average per 12 months | 5.7 | Kohn et al. (2005) |
| Anorexia nervosa | Latin America | 2003–15 | Point prevalence | 0.1 | Kolar et al. (2016) |
| Anxiety | Latin America and Caribbean | 1980–2004 | Average per 12 months | 3.4 | Kohn et al. (2005) |
| Bipolar | Latin America and Caribbean | 1980–2004 | Average per 12 months | 0.8 | Kohn et al. (2005) |
| Bulimia nervosa | Latin America | 2003–15 | Point prevalence | 1.16 | Kolar et al. (2016) |
| Binge eating | Latin America | 2003–15 | Point prevalence | 3.53 | Kolar et al. (2016) |
| Dysthymia | Latin America and Caribbean | 1980–2004 | Average per 12 months | 1.7 | Kohn et al. (2005) |
| Major depression | Latin America and Caribbean | 1980–2004 | Average per 12 months | 4.9 | Kohn et al. (2005) |
| Nonaffective psychoses | Latin America and Caribbean | 1980–2004 | Average per 12 months | 1.0 | Kohn et al. (2005) |
| Obsessive-compulsive | Latin America and Caribbean | 1980–2004 | Average per 12 months | 1.4 | Kohn et al. (2005) |
| Panic disorder | Latin America and Caribbean | 1980–2004 | Average per 12 months | 1.0 | Kohn et al. (2005) |
| **Neurological disorders** | | | | | |
| Dementia >60 years | Latin America | 1980–2009 | Age standardized | 8.5 | Prince et al. (2013) |
| Epilepsy | Latin America and Caribbean | 1982–2004 | Median Life-time prevalence | 1.8 | Burneo et al. (2005) |
| Multiple sclerosis | Latin America and Caribbean | 1995–2011 | Point prevalence | 0.02 | Cristiano et al. (2013) |
| Parkinson's disease | South America | 1985–2010 | Point prevalence | Male 1.3 Female 0.8 | Pringsheim et al. (2014) |

**FIGURE 6.1** An overview of Latin American countries and their populations.
The numbers below the country names represent their current populations (in millions of inhabitants).
*Data about ethnic groups was taken from De Blij, H.J., Muller, P.O., Nijman, J., WinklerPrins, A.M.G.A.,*
*2011. The World Today: Concepts and Regions in Geography, fifth ed. John Wiley & Sons.*

individuals from the four most populated areas in Brazil, European ancestry was the most predominant, ranging from 60.6% in the northeast to 77.7% in the south (Pena et al., 2011).

This complex knowledge of ancestry origins creates a context for the understanding of their health implications and potential health consequences (Adhikari et al., 2016; Moreno-Estrada et al., 2014; Ruiz-Linares et al., 2014). Taking into account that the entire Latin American region has experienced hundreds of years of admixture and is composed of variable rates of Caucasian, African, and Native American genes, there is a weak correlation with skin color (Parra et al., 2003). Even in global consortia data, Native American ancestry is usually underrepresented (and consequently not deeply investigated) in comparison with Caucasians, which comprise the majority of the worldwide genetic consortia (such as Hapmap or the 1000 Genomes initiatives) (Sudmant et al., 2015). High levels of Native American ancestry may have an impact on the genetic basis of NPG in the continent (De Castro and Restrepo, 2015). A recent international initiative, the Exome Aggregation Consortium, analyzed the data for about 60,000 exomes, including more than 5000 exomes from Latin American individuals (mainly from Mexico) (Lek et al., 2016).

The particular features of the genetic and environmental background of Latin American populations are known to be correlated with different risks for human diseases (including NPD) and endophenotypes; this also applies to the response of pharmacological treatments (Adhikari et al., 2016; Suarez-Kurtz and Pena, 2006). As an example of environmental factors for NPDs in Latin America, in the 2004 Pelotas Birth Cohort Study, children from low-income families had a higher prevalence of any mental disorder, especially for attention-deficit/hyperactivity disorder (ADHD), than those from wealthy families (14% vs. 8%) (Santos et al., 2014). In the same cohort, the authors evaluated the prevalence of intellectual disabilities, observing a high frequency of environmental causes, which underscored the importance of early interventions for children so as to minimize the impact of NPDs (Karam et al., 2016).

## Advances in Molecular Genetics of Neurological Diseases in Latin America

Considering that genetic factors play an important role in the susceptibility to neurological disorders (Zoghbi and Warren, 2010), several studies have analyzed candidate genes for NDs in samples from Latin American countries (Table 6.2). As non-Mendelian forms of NDs represent the large majority of cases, several studies have explored polymorphisms in candidate genes (such as *APOE*, *COMT*, and *TOMM40*, among others) that might be associated with NDs in Latin American countries, such as Alzheimer's disease, stroke, and Parkinson's disease (Arboleda et al., 2001; Benitez et al., 2010; Camelo et al., 2004; Forero et al., 2006a,b; Isordia-Salas et al., 2010; Pereira et al., 2012; Vieira et al., 2016). For ADHD, several genes (such as *SNAP25*, *SLC6A3*, and *GRM7*) have been studied (Akutagava-Martins et al., 2014a,b; Bruxel et al., 2016; Castro et al., 2013; Cupertino et al., 2017; de Azeredo et al., 2014; Fonseca et al., 2015; Gálvez et al., 2014; Genro et al., 2007; Guimaraes et al., 2009a; Guimaraes et al., 2007; Kortmann et al., 2013; Martínez-levy et al., 2013; Mota et al., 2015; Roman et al., 2003; Salatino-Oliveira et al., 2012a; Salatino-Oliveira et al., 2015; Tovo-Rodrigues et al., 2012). Other studies have analyzed mutations in causal genes (such as *PSEN1*, *PSEN2*, and

Table 6.2 An Overview of Published Association Studies for Candidate
Genes and Neurological Diseases in Latin American Countries

| Disease | Country | Genes | Reference |
|---------|---------|-------|-----------|
| Alzheimer's disease | Colombia | *APOE* | Arboleda et al. (2001) |
| Alzheimer's disease | Colombia | *A2M, ACE* | Camelo et al. (2004) |
| Alzheimer's disease | Colombia | *BDNF, COMT, UCHL1* | Forero et al. (2006b) |
| Alzheimer's disease | Colombia | *LRP1, MAPT, SLC6A4* | Forero et al. (2006a) |
| Alzheimer's disease | Brazil | *COMT* | Pereira et al. (2012) |
| Alzheimer's disease | Colombia | *PSEN1 (Mutations)* | Lalli et al. (2014) |
| Alzheimer's disease | Argentina | *PSEN2 (Mutations)* | Muchnik et al. (2015) |
| Alzheimer's disease | Brazil | *GAB2, BDNF* | Vieira et al. (2015) |
| Alzheimer's disease | Colombia | *TOMM40, CR1, PVRL2, SORL1, PICALM* | Ortega-Rojas et al. (2016) |
| Alzheimer's disease | Brazil | *DCHS2* | Vieira et al. (2016) |
| Attention deficit and hyperactivity disorder | Brazil | *ADRA2A* | Roman et al. (2003) |
| Attention deficit and hyperactivity disorder | Brazil | *SLC6A3* | Genro et al. (2007) |
| Attention deficit and hyperactivity disorder | Brazil | *SLC6A4, HTR2A* | Guimaraes et al. (2007) |
| Attention deficit and hyperactivity disorder | Brazil | *HRT1B* | Guimaraes et al. (2009a) |
| Attention deficit and hyperactivity disorder | Brazil | *DRD4* | Tovo-Rodrigues et al. (2012) |
| Attention deficit and hyperactivity disorder | Brazil | *GIT1* | Salatino-Oliveira et al. (2012a) |
| Attention deficit and hyperactivity disorder | Colombia | *ADRA2A* | Castro et al. (2013) |
| Attention deficit and hyperactivity disorder | Mexico | *COMT* | Martínez-Levy et al. (2013) |
| Attention deficit and hyperactivity disorder | Brazil | *NR3C2* | Kortmann et al. (2013) |
| Attention deficit and hyperactivity disorder | Brazil | *SLC6A3* | de Azeredo et al. (2014) |
| Attention deficit and hyperactivity disorder | Brazil | *GRM1, GRM5, GRM8 (CNVs)* | Akutagava-Martins et al. (2014b) |
| Attention deficit and hyperactivity disorder | Brazil | *GRM7* | Akutagava-Martins (2014a) |
| Attention deficit and hyperactivity disorder | Colombia | *SNAP25* | Gálvez et al. (2014) |
| Attention deficit and hyperactivity disorder | Brazil | *CDH13, CTNNA2* | Salatino-Oliveira et al. (2015) |
| Attention deficit and hyperactivity disorder | Colombia | *DRD4, DRD5, SLC6A3, DBH, HTR1B, SLC6A4* | Fonseca et al. (2015) |
| Attention deficit and hyperactivity disorder | Brazil | *NCAM1, TTC1, ANKK1, DRD2* | Mota et al. (2015) |

Table 6.2 An Overview of Published Association Studies for Candidate Genes and Neurological Diseases in Latin American Countries (*cont.*)

| Disease | Country | Genes | Reference |
|---|---|---|---|
| Attention deficit and hyperactivity disorder | Brazil | *GAD1* | Bruxel et al. (2016) |
| Attention deficit and hyperactivity disorder | Brazil | *SNAP25, NOS1* | Salatino-Oliveira et al. (2016a) |
| Attention deficit and hyperactivity disorder | Brazil | *SYT1* | Cupertino et al. (2017) |
| Ischemic stroke | Mexico | *MTHFR* | Isordia-Salas et al. (2010) |
| Parkinson's disease | Colombia | *A2M, ACE, BDNF, COMT, MAPT, SLC6A3, SLC6A4, UCHL1* | Benitez et al. (2010) |
| Parkinson's disease | Colombia | *LRRK2 (Mutations)* | Duque et al. (2015) |

*LRRK2*, among others) for hereditary forms of NDs (Duque et al., 2015; Lalli et al., 2014; Muchnik et al., 2015). In Colombia, there is the largest familial cluster of Alzheimer's disease around the world, and it is caused by a mutation in the *PSEN1* gene (p.E280A) that leads to an autosomal-dominant hereditary type of early-onset Alzheimer's disease (Sepulveda-Falla et al., 2012). Venezuela has the largest familial cluster of Huntington's disease, caused by a polyglutamine mutation in the *HTT* gene (Castilhos et al., 2016). In a small city from the northwestern of Colombia, there are around 25 multigenerational families affected with Huntington's disease (De Castro and Restrepo, 2015). These clusters of families with genetic disorders are usually the result of high levels of consanguinity, in the context of genetic isolates, which are quite useful for studying major genes (Dahdouh et al., 2016). Some of these isolated populations have been studied in genome-wide linkage and association studies (carried out in collaboration with research groups in high-income countries) focused on several NPDs, such as ADHD (Arcos-Burgos et al., 2010), bipolar disorder (Kremeyer et al., 2010), and Tourette syndrome (Scharf et al., 2013). Some epileptic syndromes, such as juvenile myoclonic epilepsy, have a well-defined genetic background. Studies have shown that mutations in EFHC1 gene are found in 9% of consecutive juvenile myoclonic epilepsy cases from neurology clinics in Mexico and Honduras (Medina et al., 2008).

There are very few studies that have analyzed DNA methylation markers for NDs in Latin America (Hernandez et al., 2014). Several meta-analyses for candidate genes and telomere length have been carried out in Latin American countries for AD, PD, and stroke (Forero et al., 2009; Forero et al., 2016a,b; Gonzalez-Giraldo et al., 2016a), in addition to computational analyses of the functional pathways for candidate genes for some of these disorders (Forero et al., 2016c; Guio-Vega and Forero, 2017).

## Advances in Molecular Genetics of Psychiatric Disorders in Latin America

Taking into account the large heritability found for several PDs (Burmeister et al., 2008) and the huge impacts on public health for some of them, multiple groups have analyzed genetic risk factors for PDs in Latin American countries (Table 6.3). As for other complex disorders, the risk for PDs is hypothesized as being the result from large numbers of variants with small effects. Many of the candidate genes and variants were chosen from pathways proposed from biochemical and pharmacological hypotheses or from regions identified in genome-wide linkage or association studies. Several research groups have explored the possible association of candidate genes (such as *SLC6A3*, *TPH2*, and *BDNF*, among others) with PDs, such as alcohol dependence, antisocial personality disorder, BP, generalized anxiety disorder, obsessive-compulsive disorder, autism, eating disorders, MDD, SZ, and suicidal behavior (Alvim-Soares et al., 2013; Bau et al., 2001; Bertola et al., 2007; Cajal et al., 2012; Campos et al., 2010, 2011; Cappi et al., 2012; Contini et al., 2012, 2006; Cordeiro et al., 2012, 2010, 2009a, 2005, 2004; Cuartas Arias et al., 2011; da Silva et al., 2016; Figueira et al., 2010; Fridman et al., 2003; González-castro et al., 2015, 2013a,b; Gonzalez et al., 2013; Gregório et al., 2005; Hernandez et al., 2016; Hounie et al., 2008; Junqueira et al., 2004; Kohlrausch et al., 2016; Longo et al., 2009; Magno et al., 2010; Marquez et al., 2013; Meira-Lima et al., 2005, 2004; Miguita et al., 2006, 2007; Moreira et al., 2015; Mota et al., 2013; Neves et al., 2011; Peralta-Leal et al., 2012; Pereira et al., 2014; Pereira Pde et al., 2011; Prestes et al., 2007; Rocha et al., 2011; Rocha et al., 2010; Segal et al., 2009; Sesarini et al., 2015, 2014; Tovilla-Zarate et al., 2013; Urraca et al., 2011; Vasconcelos et al., 2015). Several meta-analyses for candidate genes have been carried out in Latin American countries for SZ (González-castro et al., 2016), BP (González-castro et al., 2015), suicidal behavior (Arboleda et al., 2001; González-castro et al., 2013a,b), and alcohol dependence (Forero et al., 2015).

Studies in Latin America represent a small fraction of the international literature in psychiatric genetics research (Gatt et al., 2015), and some of these genetic works had as an objective the replication of previous findings from candidates identified in other populations; however, it has been common that these studies do not confirm previous findings (Gatt et al., 2015). One of the possible reasons for such inconsistencies may result from the heterogeneity of the populations or from the differences in the genes involved in the pathophysiology of these disorders (Burmeister et al., 2008).

Latin American populations provide an interesting setting to study gene-environment interactions. These countries are composed of an admixed population that lives mostly in large cities with frequent social and economic challenges (such as large numbers of inhabitants with low incomes and the presence of

**Table 6.3** An Overview of Published Association Studies for Candidate Genes and Psychiatric Disorders in Latin American Countries

| Disorder | Country | Genes | Reference |
|---|---|---|---|
| Alcohol dependence | Brazil | *DRD4, SLC6A3* | Bau et al. (2001) |
| Alcohol dependence and impulsive behaviors | Brazil | *MAOA* | Contini et al. (2006) |
| Alcohol dependence | Brazil | *HTR1B* | Contini et al. (2012) |
| Alcohol dependence | Brazil | *DRD2, DRD4* | Mota et al. (2013) |
| Alcohol dependence | Brazil | *SLC6A3, DRD2* | Vasconcelos et al. (2015) |
| Antisocial personality disorder | Colombia | *SLC6A4, HTR1A, HTR2C, HTR1B, HTR2A, SLC18A2, SNAP25, TPH1, COMT* | Cuartas Arias et al. (2011) |
| Autism spectrum disorders | Brazil | *5-HTTLPR* | Longo et al., 2009 |
| Autism | Argentina | *GABAA* | Sesarini et al. (2014) |
| Autism | Argentina | *GABRG2, GABRA4* | Sesarini et al. (2015) |
| Bipolar disorder | Brazil | *ALOX12* | Fridman et al. (2003) |
| Bipolar disorder | Brazil | *SLC6A4* | Meira-Lima et al. (2005) |
| Bipolar disorder | Brazil | *TPH2* | Campos et al. (2010) |
| Bipolar disorder | Brazil | *BDNF* | Neves et al. (2011) |
| Bipolar disorder with panic disorder comorbidity | Brazil | *TPH2* | Campos et al. (2011) |
| Bipolar disorder | Mexico, Guatemala, and Costa Rica | *CACNA1C* | Gonzalez et al. (2013) |
| Bipolar disorder | Mexico | *BDNF* | González-Castro et al. (2015) |
| Eating disorder | Mexico | *HTR1B* | Hernandez et al. (2016) |
| Generalized anxiety disorder | Brazil | *BDNF* | Moreira et al. (2015) |
| Major depressive disorder with alcohol dependence | Brazil | *GNB3* | Prestes et al. (2007) |
| Major depressive disorder with suicide attempt | Brazil | *SLC6A4* | Segal et al. (2009) |
| Major depressive disorder | Argentina | *SLC6A4* | Cajal et al. (2012) |
| Major depressive disorder | Mexico | *SLC6A4* | Peralta-Leal et al. (2012) |
| Major depressive disorder | Brazil | CRHR1 | da Silva et al. (2016) |
| Late-onset depression | Brazil | *TPH2* | Pereira Pde et al. (2011) |

**Table 6.3** An Overview of Published Association Studies for Candidate Genes and Psychiatric Disorders in Latin American Countries (*cont.*)

| Disorder | Country | Genes | Reference |
|---|---|---|---|
| Late-onset depression | Brazil | *AKT1, AKTIP* | Pereira et al. (2014) |
| Obsessive-compulsive disorder | Brazil | *COMT, SLC6A4, HTR2A* | Meira-Lima et al. (2004) |
| Obsessive-compulsive disorder | Brazil | *SLC6A2* | Miguita et al. (2006) |
| Obsessive-compulsive disorder | Brazil | *SLC6A3* | Miguita et al. (2007) |
| Obsessive-compulsive disorder | Brazil | *TNF* | Hounie (2008) |
| Obsessive-compulsive disorder | Brazil | *SLC6A4* | Rocha et al. (2009) |
| Obsessive-compulsive disorder | Brazil | *NFKBIL1* | Cordeiro et al. (2009a) |
| Obsessive-compulsive disorder | Brazil | *BDNF* | Rocha et al. (2010) |
| Obsessive-compulsive disorder | Brazil | *TPH2* | Rocha et al. (2011) |
| Obsessive-compulsive disorder | Brazil | *TNF* | Cappi et al. (2012) |
| Obsessive-compulsive disorder | Mexico | *BDNF* | Marquez et al. (2013) |
| Obsessive-compulsive disorder | Brazil | *GRIN2B* | Kohlrausch et al. (2016) |
| Postpartum depression | Brazil | *BDNF* | Figueira et al. (2010) |
| Postpartum depression | Brazil | *COMT* | Alvim-Soares et al. (2013) |
| Schizophrenia | Brazil | *PLA2G1B, KAT5, PLA2G6, PLA2G7* | Junqueira et al. (2004) |
| Schizophrenia | Brazil | *SLC6A3* | Cordeiro et al. (2004) |
| Schizophrenia and Bipolar disorder | Brazil | *RTN4* | Gregório et al. (2005) |
| Schizophrenia and Bipolar disorder | Brazil | *RGS4* | Cordeiro et al. (2005) |
| Schizophrenia | Brazil | *DRD2* | Cordeiro et al. (2009a) |
| Schizophrenia | Brazil | *HTR2A* | Bertola et al. (2007) |
| Schizophrenia | Brazil | *SLC6A3* | Cordeiro et al. (2010) |
| Schizophrenia | Mexico | *DRD3* | Urraca et al. (2011) |
| Schizophrenia | Brazil | *COMT* | Cordeiro et al. (2012) |
| Schizophrenia | Mexico | *COMT* | Tovilla-Zarate et al. (2013) |
| Suicidal behavior in bipolar patients | Brazil | *AKT1, AKTIP* | Magno et al. (2010) |
| Suicidal behavior | Mexico | *HTR1A* | González-Castro et al. (2013a) |
| Suicidal behavior | Mexico | *HTR2A* | González-Castro et al. (2013b) |

high rates of unemployment and violence) (Cia et al., 2010). These populations, which have a considerable amount of people living in environments with higher risk to develop PDs, provide a useful setting to study how genes and environmental risk factors act together to lead to vulnerability or resilience to mental illness.

There are very few studies that have analyzed epigenetic markers or gene-environment interactions for PDs in Latin America (Lima et al., 2015). Recently, telomere attrition has been studied in South American samples because of the potential relationship between telomere shortening and psychological and biological stress in PDs, such as BP and ADHD (Costa Dde et al., 2015; Lima et al., 2015), taking advantage of the potentials of studying individuals from populations with particular genetic and environmental features (Mitchell et al., 2014).

In recent years, several Latin American research groups have joined international initiatives to identify regions and genes, including CNVs, associated with some PDs, such as obsessive-compulsive disorders and Tourette syndrome (Nag et al., 2013; Scharf et al., 2013; Stewart et al., 2013).

## Molecular Genetics and Neuropsychiatric Endophenotypes in Latin America

Considering the importance of the study of endophenotypes in NPG (Flint and Munafo, 2007), several groups have researched candidate genes for intermediate phenotypes of neuropsychiatric relevance (Table 6.4). Some studies have been carried out on healthy subjects (Cruz-Fuentes et al., 2014; Forero et al., 2016a; Gonzalez-Giraldo et al., 2015a,b,c; Gonzalez-Giraldo et al., 2016b,c; Gonzalez-Giraldo et al., 2014; Ojeda et al., 2014a,b; Ojeda et al., 2013; Perea et al., 2014; Solís-ortiz et al., 2010; Speck-Hernandez et al., 2015) and others on patients diagnosed with, for example, with ADHD, BP, or SZ (Agudelo et al., 2015; Akutagava-Martins et al., 2016; Fresan et al., 2007; González-castro et al., 2015; González-castro et al., 2013b; Guimaraes et al., 2009b; Lopez-Narvaez et al., 2015; Morales-Marin et al., 2016; Salatino-Oliveira et al., 2016a,b; Salatino-Oliveira et al., 2012b; Salatino-Oliveira et al., 2015; Tovilla-Zarate et al., 2014; Tovo-Rodrigues et al., 2013; Zeni et al., 2016). These studies have mainly used validated psychological scales and tests, some of them employing computerized platforms, to determine functioning in multiple behavioral dimensions and their correlations with variants in candidate genes.

The list of candidate genes studied for neuropsychiatric endophenotypes in Latin America includes the following: BDNF, COMT, MAOA, DRD4, SLC6A4, and PER3, among others. Studies in healthy subjects have evaluated endophenotypes related with working memory, circadian rhythm, aggressiveness, and depressive symptoms. On the other hand, works in patients with neuropsychiatric diseases in Latin America have analyzed verbal working memory, suicidal

**Table 6.4** An Overview of Published Association Studies for Candidate Genes and Neuropsychiatric Endophenotypes in Latin American Countries

| Endophenotype | Country | Genes | Reference |
|---|---|---|---|
| Aggressiveness | Colombia | MIR124-1 | González-Giraldo et al. (2015a) |
| Aggression in schizophrenia | Mexico | DRD4, MAOA | Fresan et al. (2007) |
| Arithmetical processing | Colombia | BDNF, COMT | González-Giraldo et al. (2014) |
| BDNF serum levels in ADHD | Brazil | BDNF | Zeni et al. (2016) |
| Childhood adversities in adolescents | Mexico | BDNF, SLC6A4 | Cruz-Fuentes et al. (2014) |
| Daytime sleepiness and diurnal preference | Colombia | COMT, SLC6A4 | Ojeda et al. (2014a) |
| Daytime sleepiness | Colombia | MAOA | Ojeda et al. (2014b) |
| Delusions in schizophrenia | Mexico | MDR1 | Tovilla-Zarate et al. (2014) |
| Depressive symptoms | Colombia | BDNF, COMT, MAOA, SLC6A3, SLC6A4 | González-Giraldo et al. (2015b) |
| Disruptive behavior disorders in ADHD | Brazil | COMT | Salatino-Oliveira et al. (2012b) |
| Diurnal preference | Colombia | PER2, PER3 | Ojeda et al. (2013) |
| Diurnal preference | Colombia | PER3 | Perea et al. (2014) |
| Executive function and selective attention | Mexico | COMT | Solís-Ortiz et al. (2010) |
| Executive function in ADHD | Colombia | SLC6A3 | Agudelo et al. (2015) |
| Hyperactivity/inattention scores in ADHD | Brazil | DRD4 | Tovo-Rodrigues et al. (2013) |
| Hyperactivity/inattention scores in ADHD | Brazil | COMT, SLC6A3 | Akutagava-Martins et al. (2016) |
| Hyperactivity/impulsive symptoms in ADHD | Brazil | CDH13 | Salatino-Oliveira et al. (2015) |
| Obesity in bipolar disorder | Mexico | BDNF | Morales-Marin et al. (2016) |
| Oppositional symptoms in ADHD | Brazil | MAOA | Guimaraes et al. (2009b) |
| Planning performance | Colombia | PER3 | González-Giraldo et al. (2015c) |
| Suicidal behavior in bipolar patients | Mexico | BDNF | González-castro et al. (2015) |
| Suicidal behavior | Mexico | 5HTR2A | González-castro et al. (2013b) |
| Suicidal behavior | Mexico | THP2 | Lopez-Narvaez et al. (2015) |
| Stroop test performance | Colombia | MIR137 | González-Giraldo et al. (2016b) |
| Verbal working memory in ADHD | Brazil | MAP1B, NOS1 | Salatino-Oliveira et al. (2016b) |
| Verbal working memory in ADHD | Brazil | NOS1, SNAP25 | Salatino-Oliveira et al. (2016a) |
| Visual-motor tracking performance | Colombia | BDNF | González-Giraldo et al. (2016c) |

behavior, hyperactivity, aggression, and executive function. Brazil, Colombia, and Mexico have made major contributions to the knowledge about genetics and endophenotypes in Latin American samples.

## Pharmacogenetics and Neuropsychiatric Disorders in Latin America

Pharmacogenomics applied to NPDs will help to implement genomics in the care of neuropsychiatric patients in Latin America by predicting, based on the individual genetic profiles, whether a drug will treat the disease or whether a patient has a predisposition to adverse events (Suarez-Kurtz and Pena, 2006). In addition, this research will help with clinical decisions, such as the best drug and dosage for an individual, decreasing the time and costs of identifying the right pharmacological treatments. The first step in pharmacogenetics research is to identify the genetic variants that determine the response to a drug and to determine if the findings in other populations can be applied to the different samples of patients in the Latin America region (Suarez-Kurtz and Pena, 2006).

As molecular genetic factors are important variables for the response to pharmacological compounds used for the treatment of NPD, several studies on pharmacogenetics have been carried out in Latin American countries (Table 6.5), particularly in Mexico and Brazil. These studies have included ADHD, BP, MDD, OCD, and SZ patients, testing the effect of variants on candidate genes (such as *DRD1*, *SLC6A4*, and *MAOA*, among others) to the response of drugs such as methylphenidate, clozapine, and lithium (Bruxel et al., 2015; Bruxel et al., 2013; Contini et al., 2011, 2010; Corregiari et al., 2012; da Silva et al., 2008; Gonzalez-Covarrubias et al., 2016; Guimaraes et al., 2009b; Kohlrausch et al., 2008a,b; Kohlrausch et al., 2010, 2013; Michelon et al., 2006; Ota et al., 2012; Peñas-lledó et al., 2013; Polanczyk et al., 2007; Salatino-Oliveira et al., 2011; Zeni et al., 2007). Most of the studies on neurological disorders focused on epilepsy and aimed to estimate the allele and genotype frequencies of markers influencing the effects of antiepileptic drugs, evaluating if these frequencies were different between the Mexican-Mestizo populations and other reported samples (Fricke-Galindo et al., 2016). There are no studies on epigenetic markers as possible predictors of response to pharmacological treatments on NPDs in Latin America (Reynolds and Fachim, 2016).

To assess whether findings from other populations can be extrapolated to the samples from Latin America, there have been several studies and initiatives, which have estimated the frequency of the most relevant pharmacogenetic biomarkers for neuropsychiatric and metabolic phenotypes, in healthy volunteers from Latin American populations and have compared them to the frequencies observed in other samples around the world (Bonifaz-Pena et al., 2014; Cespedes-Garro et al., 2015; Rodeiro et al., 2012; Suarez-Kurtz, 2004).

Table 6.5 An Overview of Published Pharmacogenetic Studies for Neuropsychiatric Disorders in Latin American Countries

| Disease/Treatment | Country | Genes | Reference |
|---|---|---|---|
| ADHD/methylphenidate | Brazil | DRD4, SLC6A3, HTR1B, HTR2A, SLC6A4 | Zeni et al. (2007) |
| ADHD/methylphenidate | Brazil | ADRA2A | Polanczyk et al. (2007) |
| ADHD/methylphenidate | Brazil | ADRA2A | Da Silva et al. (2008) |
| ADHD/methylphenidate | Brazil | MAOA | Guimaraes et al. (2009b) |
| Adult ADHD/methylphenidate | Brazil | SLC6A3 | Contini et al. (2010) |
| ADHD/methylphenidate | Brazil | COMT | Salatino-Oliveira et al. (2011) |
| Adult ADHD/methylphenidate | Brazil | ADRA2A | Contini et al. (2011) |
| ADHD/methylphenidate | Brazil | CES1 | Bruxel et al. (2013) |
| ADHD/methylphenidate | Brazil | LPHN3 | Bruxel et al. (2015) |
| Bipolar disorder/lithium | Brazil | INPP1, SLC6A4, BDNF, TFAP2B, GSK3B | Michelon et al. (2006) |
| Major depressive disorder/ fluoxetine or amitriptyline | Mexico | CYP2D6 | Peñas-Lledó et al. (2013) |
| Obsessive-compulsive disorder/citalopram | Brazil | HTR1B, HTR2A | Corregiari et al. (2012) |
| Schizophrenia/antipsychotic treatment response | Brazil | CYP3A5, DRD3 | Kohlrausch et al. (2008a) |
| Schizophrenia/clozapine | Brazil | GNB3 | Kohlrausch et al. (2008b) |
| Schizophrenia/clozapine | Brazil | SLC6A4 | Kohlrausch et al. (2010) |
| Schizophrenia/antipsychotic treatment response | Brazil | DRD1 | Ota et al. (2012) |
| Schizophrenia/clozapine | Brazil | CYP1A2 | Kohlrausch et al. (2013) |
| Schizophrenia/haloperidol | Mexico | GSTM1 | Gonzalez-Covarrubias et al. (2016) |

There are several working groups in Latin America that have as a goal the translation of pharmacogenomics into clinical practice. The Ibero-American Network of Pharmacogenetics and Pharmacogenomics (www.ribef.com) is an initiative created in 2006 that has as its aim the promotion of personalized medicine and collaborative pharmacogenetic research in Spanish and Portuguese speaking countries. The RIBEF organization has promoted several initiatives, such as the CEIBA consortium and the MESTIFAR project. The CEIBA consortium studies the variability of phenotypes and genotypes in Latin America that are relevant to pharmacogenetics. MESTIFAR aims to analyze the ethnicity, genotype, and/or metabolic phenotype in Ibero-American populations.

In 2016 they had data for 6060 healthy volunteers (admixed, Caucasians, native Americans, Jews, and Afro-descendants) (Sosa-Macias et al., 2016). Other international initiatives that are carrying out pharmacogenetic research and promoting pharmacogenetics in Latin America are the Pharmacogenetics for Every Nation Initiative (PGENI) (http://www.pgeni.org) and The Golden Helix Institute of Biomedical Research (http://goldenhelix.org).

Most of the regulatory agencies in Latin America do not enforce specific regulations for conducting pharmacogenetic research. Regulations may be inferred from general norms of clinical research and international regulations. In the case of new drugs, sponsors that perform pharmacogenetic studies are required to gain regulatory approval, and specific guidelines for pharmacogenetics evaluation is needed in most Latin American countries (Quinones et al., 2014). Other international resources of interest are the Pharmacogenomics Research Network (PGRN) (www.pgrn.org), the Clinical Pharmacogenetics Implementation Consortium (CPIC) (www.pharmgkb.org/page/cpic), and the Pharmacogenomics Knowledge Base PharmGKB www.pharmgkb.org.

## Perspectives for Future Studies

Several challenges remain for a broad implementation of genomic medicine into the care of NPDs in Latin American countries. In the last 20 years, we have learned that NPDs are complex and multifactorial entities, with the potential effect of hundreds of genetic and epigenetic factors interacting with multiple environmental variables, leading to particular individual susceptibilities to specific diseases (Burmeister et al., 2008).

In Latin America, because of historical reasons, research initiatives into NPDs are usually not included in the top national priorities, leading to lack of adequate funding for the organization of the necessary clinical services and research infrastructure. There is the need for larger public and private funding for research into NPG in Latin American countries (Sharan et al., 2009). For this increase in funding, it will be fundamental that researchers develop more activities aimed to create a deeper awareness of the importance of research into NPG among the general public, regional and national governments, professional and academic societies, and other key stakeholders. Data about prevalence, heritability, and burden of NPDs will help to show that they should become key priorities for research in Latin America (Forero et al., 2014). Most of the epidemiological studies performed in Latin America have been cross-sectional works that did not include subsequent follow ups. The development of longitudinal cohort studies in the general population will open new avenues for research, both nationally and internationally.

Collaborations between laboratories are fundamental for research projects in NPG and are vital for a deeper understanding of the pathophysiology of

NPDs and for developing better treatment strategies for the affected patients. Those collaborations might be between labs from Latin America and from developed countries (North America and Europe, among others) and between Latin American research groups. As global research into NPG is moving to large interinstitutional and international consortia, there is the need for the creation of Latin American networks for NPG, which could be for specific diseases or for general research initiatives (Mitropoulos et al., 2015).

The strategic implementation of genomic technologies is a fundamental need in Latin American countries, given the importance of the local development of novel experimental methods (for diminishing the dependence on collaborations with foreign groups), such as whole exome and genome sequencing, and genome-wide expression and DNA methylation analyses (Forero et al., 2016d). A collaboration of several research groups and institutions might be key for these initiatives that need large funding from public or private agencies and a strong bioinformatics structure (Forero et al., 2016d).

Given the complexity of the clinical diagnosis of NPD, particularly in Latin American health institutions, there is an urgent need for the existence of an adequate number of neurologists, psychiatrists, and psychologists with appropriate research training and experience (Forero et al., 2014). In this context, there is the need for the local validation of additional psychosocial and clinical scales and for the existence of neuroimaging and neurophysiological platforms. More studies about the genetic basis of neuropsychiatric endophenotypes, in both healthy subjects and patients, are needed in Latin America. An improvement of technological infrastructure applied to health care is essential, considering that several Latin American countries have large territorial areas, facilitating a broader availability of health assistance and a better quality of health care.

In relation to pharmacogenetics, there is a need to continue the identification of genetic variants that determine the response to drugs in Latin American populations. An important step would be to join and establish working groups and consortia (https://www.pharmgkb.org/page/collaborators and www.ribef.com). With time, the costs of genetic lab tests and kits will come down and they will become more affordable and common. Other countries (Canada, Europe, and Japan) are including genetic information related to neuropsychiatric disorders in drug labels. There are more than 30 psychiatric drugs that contain genetic information on their labels, and several drugs require testing before prescribing the medication (https://www.pharmgkb.org/view/drug-labels.do). In Latin America, the need is for efforts aimed to create detailed guidelines for the implementation of pharmacogenomics, as well as for education on awareness about pharmacogenomics among clinicians and regulatory institutions, in order to facilitate the use of pharmacogenetics tests (Quinones et al., 2014).

As reviewed in previous sections, there are several diseases of high epidemiological impact that have been relatively understudied in Latin American countries (e.g., autism, MDD, migraine, multiple sclerosis, epilepsies, and alcohol and drug abuse). A focus on the large burden of these disorders might benefit from the possibility that public agencies would provide available funds for its mitigation (Forero et al., 2014).

## Acknowledgments

Research in the Laboratory of Neuropsychiatric Genetics (Universidad Antonio Nariño) has been supported by grants from VCTI-UAN and Colciencias (to DAF). YG-G is supported by a PhD fellowship from Centro de Estudios Interdisciplinarios Básicos y Aplicados CEIBA (Rodolfo Llinás Program). We thank Leon Ruiter Lopez for his help for the generation of Fig. 6.1.

## References

Adhikari, K., Mendoza-Revilla, J., Chacon-Duque, J.C., Fuentes-Guajardo, M., Ruiz-Linares, A., 2016. Admixture in Latin America. Curr. Opin. Genet. Dev. 41, 106–114. doi: 10.1016/j.gde.2016.09.003.

Agudelo, J.A., Gálvez, J.M., Fonseca, D.J., Mateus, H.E., Talero-Gutierrez, C., Velez-Van-Meerbeke, A., 2015. Evidence of an association between 10/10 genotype of DAT1 and endophenotypes of attention deficit/hyperactivity disorder. Neurologia 30 (3), 137–143. doi: 10.1016/j.nrl.2013.12.005.

Akutagava-Martins, G.C., Salatino-Oliveira, A., Bruxel, E.M., Genro, J.P., Mota, N.R., Polanczyk, G.V., Hutz, M.H., 2014a. Lack of association between the GRM7 gene and attention deficit hyperactivity disorder. Psychiatr. Genet. 24 (6), 281–282. doi: 10.1097/ypg.0000000000000059.

Akutagava-Martins, G.C., Salatino-Oliveira, A., Genro, J.P., Contini, V., Polanczyk, G., Zeni, C., Hutz, M.H., 2014b. Glutamatergic copy number variants and their role in attention-deficit/hyperactivity disorder. Am. J. Med. Genet. B Neuropsychiatr. Genet. 165B (6), 502–509. doi: 10.1002/ajmg.b.32253.

Akutagava-Martins, G.C., Salatino-Oliveira, A., Kieling, C., Genro, J.P., Polanczyk, G.V., Anselmi, L., Hutz, M.H., 2016. COMT and DAT1 genes are associated with hyperactivity and inattention traits in the 1993 Pelotas Birth Cohort: evidence of sex-specific combined effect. J. Psychiatry Neurosci. 41 (6), 405–412. doi: 10.1503/jpn.150270.

Alvim-Soares, A., Miranda, D., Campos, S.B., Figueira, P., Romano-Silva, M.A., Correa, H., 2013. Postpartum depression symptoms associated with Val158Met COMT polymorphism. Arch. Womens Ment. Health 16 (4), 339–340. doi: 10.1007/s00737-013-0349-8.

Arboleda, G.H., Yunis, J.J., Pardo, R., Gomez, C.M., Hedmont, D., Arango, G., Arboleda, H., 2001. Apolipoprotein E genotyping in a sample of Colombian patients with Alzheimer's disease. Neurosci. Lett. 305 (2), 135–138.

Arcos-Burgos, M., Jain, M., Acosta, M.T., Shively, S., Stanescu, H., Wallis, D., Muenke, M., 2010. A common variant of the latrophilin 3 gene, LPHN3, confers susceptibility to ADHD and predicts effectiveness of stimulant medication. Mol. Psychiatry 15 (11), 1053–1066. doi: 10.1038/mp.2010.6.

Bau, C.H., Almeida, S., Costa, F.T., Garcia, C.E., Elias, E.P., Ponso, A.C., Hutz, M.H., 2001. DRD4 and DAT1 as modifying genes in alcoholism: interaction with novelty seeking on level of alcohol consumption. Mol. Psychiatry 6 (1), 7–9.

Benitez, B.A., Forero, D.A., Arboleda, G.H., Granados, L.A., Yunis, J.J., Fernandez, W., Arboleda, H., 2010. Exploration of genetic susceptibility factors for Parkinson's disease in a South American sample. J. Genet. 89 (2), 229–232.

Bertola, V., Cordeiro, Q., Zung, S., Miracca, E.C., Vallada, H., 2007. Association analysis between the C516T polymorphism in the 5-HT2A receptor gene and schizophrenia. Arq Neuropsiquiatr. 65 (1), 11–14.

Bonifaz-Pena, V., Contreras, A.V., Struchiner, C.J., Roela, R.A., Furuya-Mazzotti, T.K., Chammas, R., Suarez-Kurtz, G., 2014. Exploring the distribution of genetic markers of pharmacogenomics relevance in Brazilian and Mexican populations. PLoS One 9 (11), e112640, 10.1371/journal.pone.0112640.

Bruxel, E.M., Akutagava-Martins, G.C., Salatino-Oliveira, A., Genro, J.P., Zeni, C.P., Polanczyk, G.V., Hutz, M.H., 2016. GAD1 gene polymorphisms are associated with hyperactivity in attention-deficit/hyperactivity disorder. Am. J. Med. Genet. B Neuropsychiatr. Genet. 171 (8), 1099–1104, 10.1002/ajmg.b.32489.

Bruxel, E.M., Salatino-Oliveira, A., Akutagava-Martins, G.C., Tovo-Rodrigues, L., Genro, J.P., Zeni, C.P., Hutz, M.H., 2015. LPHN3 and attention-deficit/hyperactivity disorder: a susceptibility and pharmacogenetic study. Genes Brain Behav. 14 (5), 419–427, 10.1111/gbb.12224.

Bruxel, E.M., Salatino-Oliveira, A., Genro, J.P., Zeni, C.P., Polanczyk, G.V., Chazan, R., Hutz, M.H., 2013. Association of a carboxylesterase 1 polymorphism with appetite reduction in children and adolescents with attention-deficit/hyperactivity disorder treated with methylphenidate. Pharmacogenomics J. 13 (5), 476–480, 10.1038/tpj.2012.25.

Burmeister, M., McInnis, M.G., Zollner, S., 2008. Psychiatric genetics: progress amid controversy. Nat. Rev. Genet. 9 (7), 527–540. doi: 10.1038/nrg2381.

Burneo, J.G., Tellez-Zenteno, J., Wiebe, S., 2005. Understanding the burden of epilepsy in Latin America: a systematic review of its prevalence and incidence. Epilepsy Res. 66 (1–3), 63–74. doi: 10.1016/j.eplepsyres.2005.07.002.

Cajal, A.R., Redal, M.A., Costa, L.D., Lesik, L.A., Faccioli, J.L., Finkelsztein, C.A., Argibay, P.F., 2012. Influence of 5-HTTLPR and 5-HTTVNTR polymorphisms of the serotonin transporter gene (SLC6A4) on major depressive disorder in a sample of Argentinean population. Psychiatr. Genet. 22 (2), 103–104. doi: 10.1097/YPG.0b013e32834acc9b.

Camelo, D., Arboleda, G., Yunis, J.J., Pardo, R., Arango, G., Solano, E., Arboleda, H., 2004. Angiotensin-converting enzyme and alpha-2-macroglobulin gene polymorphisms are not associated with Alzheimer's disease in Colombian patients. J. Neurol. Sci. 218 (1–2), 47–51. doi: 10.1016/j.jns.2003.10.008.

Campos, S.B., Miranda, D.M., Souza, B.R., Pereira, P.A., Neves, F.S., Bicalho, M.A., Correa, H., 2010. Association of polymorphisms of the tryptophan hydroxylase 2 gene with risk for bipolar disorder or suicidal behavior. J. Psychiatr. Res. 44 (5), 271–274. doi: 10.1016/j.jpsychires.2009.09.007.

Campos, S.B., Miranda, D.M., Souza, B.R., Pereira, P.A., Neves, F.S., Tramontina, J., Correa, H., 2011. Association study of tryptophan hydroxylase 2 gene polymorphisms in bipolar disorder patients with panic disorder comorbidity. Psychiatr. Genet. 21 (2), 106–111. doi: 10.1097/YPG.0b013e328341a3a8.

Cappi, C., Muniz, R.K., Sampaio, A.S., Cordeiro, Q., Brentani, H., Palacios, S.A., Hounie, A.G., 2012. Association study between functional polymorphisms in the TNF-alpha gene and obsessive-compulsive disorder. Arq Neuropsiquiatr. 70 (2), 87–90.

Castilhos, R.M., Augustin, M.C., Santos, J.A., Perandones, C., Saraiva-Pereira, M.L., Jardim, L.B., 2016. Rede Neurogenetica Genetic aspects of Huntington's disease in Latin America. A systematic review. Clin. Genet. 89 (3), 295–303. doi: 10.1111/cge.12641.

Castro, T., Mateus, H.E., Fonseca, D.J., Forero, D., Restrepo, C.M., Talero, C., Laissue, P., 2013. Sequence analysis of the ADRA2A coding region in children affected by attention deficit hyperactivity disorder. Neurol. Sci. 34 (12), 2219–2222. doi: 10.1007/s10072-013-1569-4.

Cespedes-Garro, C., Naranjo, M.E., Ramirez, R., Serrano, V., Farinas, H., Barrantes, R., 2015. Pharmacogenomics Ribef Pharmacogenetics in Central American healthy volunteers: interethnic variability. Drug Metab. Pers. Ther. 30 (1), 19–31. doi: 10.1515/dmdi-2014-0025.

Cia, A.H., Rojas, R.C., Adad, M.A., 2010. Current clinical advances and future perspectives in the psychiatry/mental health field of Latin America. Int. Rev. Psychiatry 22 (4), 340–346. doi: 10.3109/09540261.2010.501167.

Contini, V., Bertuzzi, G.P., Polina, E.R., Hunemeier, T., Hendler, E.M., Hutz, M.H., Bau, C.H., 2012. A haplotype analysis is consistent with the role of functional HTR1B variants in alcohol dependence. Drug Alcohol Depend. 122 (1–2), 100–104. doi: 10.1016/j.drugalcdep.2011.09.020.

Contini, V., Marques, F.Z., Garcia, C.E., Hutz, M.H., Bau, C.H., 2006. MAOA-uVNTR polymorphism in a Brazilian sample: further support for the association with impulsive behaviors and alcohol dependence. Am. J. Med. Genet. B Neuropsychiatr. Genet. 141B (3), 305–308. doi: 10.1002/ajmg.b.30290.

Contini, V., Victor, M.M., Cerqueira, C.C., Polina, E.R., Grevet, E.H., Salgado, C.A., Bau, C.H., 2011. Adrenergic alpha2A receptor gene is not associated with methylphenidate response in adults with ADHD. Eur. Arch. Psychiatry Clin. Neurosci. 261 (3), 205–211. doi: 10.1007/s00406-010-0172-4.

Contini, V., Victor, M.M., Marques, F.Z., Bertuzzi, G.P., Salgado, C.A., Silva, K.L., Bau, C.H., 2010. Response to methylphenidate is not influenced by DAT1 polymorphisms in a sample of Brazilian adult patients with ADHD. J. Neural Transm. (Vienna) 117 (2), 269–276. doi: 10.1007/s00702-009-0362-2.

Cordeiro, Q., Silva, R.T., Vallada, H., 2012. Association study between the rs165599 catechol-O-methyltransferase genetic polymorphism and schizophrenia in a Brazilian sample. Arq Neuropsiquiatr. 70 (12), 913–916.

Cordeiro, Q., Siqueira-Roberto, J., Vallada, H., 2010. Association between the SLC6A3 A1343G polymorphism and schizophrenia. Arq Neuropsiquiatr. 68 (5), 716–719.

Cordeiro, Q., Siqueira-Roberto, J., Zung, S., Vallada, H., 2009a. Association between the DRD2-141C insertion/deletion polymorphism and schizophrenia. Arq Neuropsiquiatr. 67 (2A), 191–194.

Cordeiro, Q., Souza, B.R., Correa, H., Guindalini, C., Hutz, M.H., Vallada, H., Romano-Silva, M.A., 2009b. A review of psychiatric genetics research in the Brazilian population. Rev. Bras. Psiquiatr. 31 (2), 154–162.

Cordeiro, Q., Talkowski, M.E., Chowdari, K.V., Wood, J., Nimgaonkar, V., Vallada, H., 2005. Association and linkage analysis of RGS4 polymorphisms with schizophrenia and bipolar disorder in Brazil. Genes Brain Behav. 4 (1), 45–50. doi: 10.1111/j.1601-183x.2004.00096.x.

Cordeiro, Q., Talkowski, M., Wood, J., Ikenaga, E., Vallada, H., 2004. Lack of association between VNTR polymorphism of dopamine transporter gene (SLC6A3) and schizophrenia in a Brazilian sample. Arq Neuropsiquiatr. 62 (4), 973–976, /S0004-282x2004000600008.

Corregiari, F.M., Bernik, M., Cordeiro, Q., Vallada, H., 2012. Endophenotypes and serotonergic polymorphisms associated with treatment response in obsessive-compulsive disorder. Clinics (Sao Paulo) 67 (4), 335–340.

Costa Dde, S., Rosa, D.V., Barros, A.G., Romano-Silva, M.A., Malloy-Diniz, L.F., Mattos, P., de Miranda, D.M., 2015. Telomere length is highly inherited and associated with hyperactivity-impulsivity in children with attention deficit/hyperactivity disorder. Front. Mol. Neurosci. 8 (28)doi: 10.3389/fnmol.2015.00028.

Cristiano, E., Rojas, J., Romano, M., Frider, N., Machnicki, G., Giunta, D., Correale, J., 2013. The epidemiology of multiple sclerosis in Latin America and the Caribbean: a systematic review. Mult. Scler. 19 (7), 844–854. doi: 10.1177/1352458512462918.

Cruz-Fuentes, C.S., Benjet, C., Martínez-levy, G.A., Perez-Molina, A., Briones-Velasco, M., Suarez-Gonzalez, J., 2014. BDNF Met66 modulates the cumulative effect of psychosocial childhood adversities on major depression in adolescents. Brain Behav. 4 (2), 290–297. doi: 10.1002/brb3.220.

Cuartas Arias, J.M., Palacio Acosta, C.A., Valencia, J.G., Montoya, G.J., Arango Viana, J.C., Nieto, O.C., Ruiz-Linares, A., 2011. Exploring epistasis in candidate genes for antisocial personality disorder. Psychiatr. Genet. 21 (3), 115–124. doi: 10.1097/YPG.0b013e3283437175.

Cupertino, R.B., Schuch, J.B., Bandeira, C.E., da Silva, B.S., Rovaris, D.L., Kappel, D.B., Mota, N.R., 2017. Replicated association of Synaptotagmin (SYT1) with ADHD and its broader influence in externalizing behaviors. Eur. Neuropsychopharmacol.doi: 10.1016/j.euroneuro.2017.01.007.

da Silva, B.S., Rovaris, D.L., Schuch, J.B., Mota, N.R., Cupertino, R.B., Aroche, A.P., Bau, C.H., 2016. Effects of corticotropin-releasing hormone receptor 1 SNPs on major depressive disorder are influenced by sex and smoking status. J. Affect Disord. 205, 282–288. doi: 10.1016/j.jad.2016.08.008.

da Silva, T.L., Pianca, T.G., Roman, T., Hutz, M.H., Faraone, S.V., Schmitz, M., Rohde, L.A., 2008. Adrenergic alpha2A receptor gene and response to methylphenidate in attention-deficit/hyperactivity disorder-predominantly inattentive type. J. Neural Transm. (Vienna) 115 (2), 341–345. doi: 10.1007/s00702-007-0835-0.

Dahdouh, A., Taleb, M., Blecha, L., Benyamina, A., 2016. Genetics and psychotic disorders: a fresh look at consanguinity. Eur. J. Med. Genet. 59 (2), 104–110. doi: 10.1016/j.ejmg.2015.12.010.

de Azeredo, L.A., Rovaris, D.L., Mota, N.R., Polina, E.R., Marques, F.Z., Contini, V., Bau, C.H., 2014. Further evidence for the association between a polymorphism in the promoter region of SLC6A3/DAT1 and ADHD: findings from a sample of adults. Eur. Arch. Psychiatry Clin. Neurosci. 264 (5), 401–408. doi: 10.1007/s00406-014-0486-8.

De Castro, M., Restrepo, C.M., 2015. Genetics and genomic medicine in Colombia. Mol. Genet. Genomic Med. 3 (2), 84–91. doi: 10.1002/mgg3.139.

Demyttenaere, K., Bruffaerts, R., Posada-Villa, J., Gasquet, I., Kovess, V., Lepine, J.P., Consortium, W.H.O. World Mental Health Survey, 2004. Prevalence, severity, and unmet need for treatment of mental disorders in the World Health Organization World Mental Health Surveys. JAMA 291 (21), 2581–2590. doi: 10.1001/jama.291.21.2581.

Disease, G.B.D., Injury, Incidence, & Prevalence, Collaborators, 2016. Global, regional, and national incidence, prevalence, and years lived with disability for 310 diseases and injuries 1990–2015: a systematic analysis for the Global Burden of Disease Study 2015. Lancet 388 (10053), 1545–1602. doi: 10.1016/S0140-6736(16)31678-6.

Duque, A.F., Lopez, J.C., Benitez, B., Hernandez, H., Yunis, J.J., Fernandez, W., Arboleda, G., 2015. Analysis of the LRRK2 p.G2019S mutation in Colombian Parkinson's Disease Patients. Colomb. Med. (Cali) 46 (3), 117–121.

Figueira, P., Malloy-Diniz, L., Campos, S.B., Miranda, D.M., Romano-Silva, M.A., De Marco, L., Correa, H., 2010. An association study between the Val66Met polymorphism of the BDNF gene and postpartum depression. Arch. Womens Ment. Health 13 (3), 285–289. doi: 10.1007/s00737-010-0146-6.

Flint, J., Munafo, M.R., 2007. The endophenotype concept in psychiatric genetics. Psychol. Med. 37 (2), 163–180. doi: 10.1017/S0033291706008750.

Fonseca, D.J., Mateus, H.E., Gálvez, J.M., Forero, D.A., Talero-Gutierrez, C., Velez-van-Meerbeke, A., 2015. Lack of association of polymorphisms in six candidate genes in colombian adhd patients. Ann. Neurosci. 22 (4), 217–221. doi: 10.5214/ans.0972.7531.220405.

Forero, D.A., Arboleda, G.H., Vasquez, R., Arboleda, H., 2009. Candidate genes involved in neural plasticity and the risk for attention-deficit hyperactivity disorder: a meta-analysis of 8 common variants. J. Psychiatry Neurosci. 34 (5), 361–366.

Forero, D.A., Arboleda, G., Yunis, J.J., Pardo, R., Arboleda, H., 2006a. Association study of polymorphisms in LRP1, tau and 5-HTT genes and Alzheimer's disease in a sample of Colombian patients. J. Neural Transm. (Vienna) 113 (9), 1253–1262. doi: 10.1007/s00702-005-0388-z.

Forero, D.A., Benitez, B., Arboleda, G., Yunis, J.J., Pardo, R., Arboleda, H., 2006b. Analysis of functional polymorphisms in three synaptic plasticity-related genes (BDNF COMT AND UCHL1) in Alzheimer's disease in Colombia. Neurosci. Res. 55 (3), 334–341. doi: 10.1016/j.neures.2006.04.006.

Forero, D.A., Gonzalez-Giraldo, Y., Lopez-Quintero, C., Castro-Vega, L.J., Barreto, G.E., Perry, G., 2016a. Meta-analysis of telomere length in Alzheimer's Disease. J. Gerontol. A Biol. Sci. Med. Sci. 71 (8), 1069–1073. doi: 10.1093/gerona/glw053.

Forero, D.A., Gonzalez-Giraldo, Y., Lopez-Quintero, C., Castro-Vega, L.J., Barreto, G.E., Perry, G., 2016b. Telomere length in Parkinson's disease: a meta-analysis. Exp. Gerontol. 75, 53–55. doi: 10.1016/j.exger.2016.01.002.

Forero, D.A., Lopez-Leon, S., Shin, H.D., Park, B.L., Kim, D.J., 2015. Meta-analysis of six genes (BDNF, DRD1, DRD3, DRD4 GRIN2B and MAOA) involved in neuroplasticity and the risk for alcohol dependence. Drug Alcohol Depend 149, 259–263. doi: 10.1016/j.drugalcdep.2015.01.017.

Forero, D.A., Prada, C.F., Perry, G., 2016c. Functional and genomic features of human genes mutated in neuropsychiatric disorders. Open Neurol. J. 10, 143–148. doi: 10.2174/1874205X01610010143.

Forero, D.A., Velez-van-Meerbeke, A., Deshpande, S.N., Nicolini, H., Perry, G., 2014. Neuropsychiatric genetics in developing countries: current challenges. World J. Psychiatry 4 (4), 69–71. doi: 10.5498/wjp.v4.i4.69.

Forero, D.A., Wonkam, A., Wang, W., Laissue, P., Lopez-Correa, C., Fernandez-Lopez, J.C., Perry, G., 2016d. Current needs for human and medical genomics research infrastructure in low and middle income countries. J. Med. Genet. 53 (7), 438–440. doi: 10.1136/jmedgenet-2015-103631.

Fresan, A., Camarena, B., Apiquian, R., Aguilar, A., Urraca, N., Nicolini, H., 2007. Association study of MAO-A and DRD4 genes in schizophrenic patients with aggressive behavior. Neuropsychobiology 55 (3–4), 171–175. doi: 10.1159/000106477.

Fricke-Galindo, I., Ortega-Vazquez, A., Monroy-Jaramillo, N., Dorado, P., Jung-Cook, H., Peñaslledó, E., Lopez-Lopez, M., 2016. Allele and genotype frequencies of genes relevant to anti-epileptic drug therapy in Mexican-Mestizo healthy volunteers. Pharmacogenomics.doi: 10.2217/pgs-2016-0078.

Fridman, C., Ojopi, E.P., Gregório, S.P., Ikenaga, E.H., Moreno, D.H., Demetrio, F.N., Dias Neto, E., 2003. Association of a new polymorphism in ALOX12 gene with bipolar disorder. Eur. Arch. Psychiatry Clin. Neurosci. 253 (1), 40–43. doi: 10.1007/s00406-003-0404-y.

Gálvez, J.M., Forero, D.A., Fonseca, D.J., Mateus, H.E., Talero-Gutierrez, C., Velez-van-Meerbeke, A., 2014. Evidence of association between SNAP25 gene and attention deficit hyperactivity disorder in a Latin American sample. Atten. Defic. Hyperact Disord 6 (1), 19–23. doi: 10.1007/s12402-013-0123-9.

Gatt, J.M., Burton, K.L., Williams, L.M., Schofield, P.R., 2015. Specific and common genes implicated across major mental disorders: a review of meta-analysis studies. J. Psychiatr. Res. 60, 1–13. doi: 10.1016/j.jpsychires.2014.09.014.

Genro, J.P., Zeni, C., Polanczyk, G.V., Roman, T., Rohde, L.A., Hutz, M.H., 2007. A promoter polymorphism (−839 C > T) at the dopamine transporter gene is associated with attention deficit/

hyperactivity disorder in Brazilian children. Am. J. Med. Genet. B Neuropsychiatr. Genet. 144B (2), 215–219. doi: 10.1002/ajmg.b.30428.

González-castro, T.B., Hernandez-Diaz, Y., Juarez-Rojop, I.E., Lopez-Narvaez, M.L., Tovilla-Zarate, C.A., Fresan, A., 2016. The role of a catechol-O-methyltransferase (COMT) Val158Met genetic polymorphism in schizophrenia: a systematic review and updated meta-analysis on 32,816 subjects. Neuromol. Med. 18 (2), 216–231. doi: 10.1007/s12017-016-8392-z.

González-castro, T.B., Nicolini, H., Lanzagorta, N., Lopez-Narvaez, L., Genis, A., Pool Garcia, S., Tovilla-Zarate, C.A., 2015. The role of brain-derived neurotrophic factor (BDNF) Val66Met genetic polymorphism in bipolar disorder: a case-control study comorbidities, and meta-analysis of 16,786 subjects. Bipolar Disord. 17 (1), 27–38. doi: 10.1111/bdi.12227.

González-castro, T.B., Tovilla-Zarate, C.A., Juarez-Rojop, I., Pool Garcia, S., Genis, A., Nicolini, H., Lopez Narvaez, L., 2013a. Association of 5HTR1A gene variants with suicidal behavior: case-control study and updated meta-analysis. J. Psychiatr. Res. 47 (11), 1665–1672. doi: 10.1016/j.jpsychires.2013.04.011.

González-castro, T.B., Tovilla-Zarate, C., Juarez-Rojop, I., Pool Garcia, S., Velazquez-Sanchez, M.P., Genis, A., Lopez Narvaez, L., 2013b. Association of the 5HTR2A gene with suicidal behavior: case-control study and updated meta-analysis. BMC Psychiatry 13, 25. doi: 10.1186/1471-244x-13-25.

Gonzalez-Covarrubias, V., Martinez-Magana, J.J., Coronado-Sosa, R., Villegas-Torres, B., Genis-Mendoza, A.D., Canales-Herrerias, P., Soberon, X., 2016. Exploring variation in known pharmacogenetic variants and its association with drug response in different mexican populations. Pharm. Res. 33 (11), 2644–2652. doi: 10.1007/s11095-016-1990-5.

Gonzalez-Giraldo, Y., Barreto, G.E., Fava, C., Forero, D.A., 2016a. Ischemic stroke and six genetic variants in CRP, EPHX2, FGA, and NOTCH3 genes: a meta-analysis. J. Stroke Cerebrovasc. Dis. 25 (9), 2284–2289. doi: 10.1016/j.jstrokecerebrovasdis.2016.05.020.

Gonzalez-Giraldo, Y., Camargo, A., Lopez-Leon, S., Adan, A., Forero, D.A., 2015a. A functional SNP in MIR124-1, a brain expressed miRNA gene, is associated with aggressiveness in a Colombian sample. Eur. Psychiatry 30 (4), 499–503. doi: 10.1016/j.eurpsy.2015.03.002.

Gonzalez-Giraldo, Y., Camargo, A., Lopez-Leon, S., Forero, D.A., 2015b. No association of BDNF, COMT, MAOA, SLC6A3, and SLC6A4 genes and depressive symptoms in a sample of healthy Colombian subjects. Depress Res. Treat. 2015, 145483. doi: 10.1155/2015/145483.

Gonzalez-Giraldo, Y., Gonzalez-Reyes, R.E., Forero, D.A., 2016b. A functional variant in MIR137, a candidate gene for schizophrenia, affects Stroop test performance in young adults. Psychiatry Res. 236, 202–205. doi: 10.1016/j.psychres.2016.01.006.

Gonzalez-Giraldo, Y., Gonzalez-Reyes, R.E., Mueller, S.T., Piper, B.J., Adan, A., Forero, D.A., 2015c. Differences in planning performance, a neurocognitive endophenotype, are associated with a functional variant in PER3 gene. Chronobiol. Int. 32 (5), 591–595. doi: 10.3109/07420528.2015.1014096.

Gonzalez-Giraldo, Y., Rojas, J., Mueller, S.T., Piper, B.J., Adan, A., Forero, D.A., 2016c. BDNF Val-66Met is associated with performance in a computerized visual-motor tracking test in healthy adults. Motor Control 20 (1), 122–134. doi: 10.1123/mc.2014-0075.

Gonzalez-Giraldo, Y., Rojas, J., Novoa, P., Mueller, S.T., Piper, B.J., Adan, A., Forero, D.A., 2014. Functional polymorphisms in BDNF and COMT genes are associated with objective differences in arithmetical functioning in a sample of young adults. Neuropsychobiology 70 (3), 152–157, 10.1159/000366483.

Gonzalez, S., Xu, C., Ramirez, M., Zavala, J., Armas, R., Contreras, S.A., Escamilla, M., 2013. Suggestive evidence for association between L-type voltage-gated calcium channel (CACNA1C) gene haplotypes and bipolar disorder in Latinos: a family-based association study. Bipolar Disord. 15 (2), 206–214. doi: 10.1111/bdi.12041.

Gregório, S.P., Mury, F.B., Ojopi, E.B., Sallet, P.C., Moreno, D.H., Yacubian, J., Dias-Neto, E., 2005. Nogo CAA 3'UTR Insertion polymorphism is not associated with Schizophrenia nor with bipolar disorder. Schizophr. Res. 75 (1), 5–9. doi: 10.1016/j.schres.2004.11.010.

Guerra, M., Ferri, C.P., Sosa, A.L., Salas, A., Gaona, C., Gonzales, V., Prince, M., 2009. Late-life depression in Peru, Mexico and Venezuela: the 10/66 population-based study. Br. J. Psychiatry 195 (6), 510–515. doi: 10.1192/bjp.bp.109.064055.

Guimaraes, A.P., Schmitz, M., Polanczyk, G.V., Zeni, C., Genro, J., Roman, T., Hutz, M.H., 2009a. Further evidence for the association between attention deficit/hyperactivity disorder and the serotonin receptor 1B gene. J. Neural Transm. (Vienna) 116 (12), 1675–1680. doi: 10.1007/s00702-009-0305-y.

Guimaraes, A.P., Zeni, C., Polanczyk, G., Genro, J.P., Roman, T., Rohde, L.A., Hutz, M.H., 2009b. MAOA is associated with methylphenidate improvement of oppositional symptoms in boys with attention deficit hyperactivity disorder. Int. J. Neuropsychopharmacol. 12 (5), 709–714. doi: 10.1017/s1461145709000212.

Guimaraes, A.P., Zeni, C., Polanczyk, G.V., Genro, J.P., Roman, T., Rohde, L.A., Hutz, M.H., 2007. Serotonin genes and attention deficit/hyperactivity disorder in a Brazilian sample: preferential transmission of the HTR2A 452His allele to affected boys. Am. J Med. Genet. B Neuropsychiatr. Genet. 144B (1), 69–73. doi: 10.1002/ajmg.b.30400.

Guio-Vega, G.P., Forero, D.A., 2017. Functional genomics of candidate genes derived from genome-wide association studies for five common neurological diseases. Int. J. Neurosci. 127 (2), 118–123. doi: 10.3109/00207454.2016.1149172.

Hernandez, H.G., Mahecha, M.F., Mejia, A., Arboleda, H., Forero, D.A., 2014. Global long interspersed nuclear element 1 DNA methylation in a Colombian sample of patients with late-onset Alzheimer's disease. Am. J. Alzheimers Dis. Other Demen. 29 (1), 50–53. doi: 10.1177/1533317513505132.

Hernandez, S., Camarena, B., Gonzalez, L., Caballero, A., Flores, G., Aguilar, A., 2016. A family-based association study of the HTR1B gene in eating disorders. Rev. Bras. Psiquiatr 38 (3), 239–242. doi: 10.1590/1516-4446-2016-1936.

Hounie, A.G., Cappi, C., Cordeiro, Q., Sampaio, A.S., Moraes, I., Rosario, M.C., Miguel, E.C., 2008. TNF-alpha polymorphisms are associated with obsessive-compulsive disorder. Neurosci. Lett. 442 (2), 86–90. doi: 10.1016/j.neulet.2008.07.022.

Isordia-Salas, I., Barinagarrementeria-Aldatz, F., Leanos-Miranda, A., Borrayo-Sanchez, G., Vela-Ojeda, J., Garcia-Chavez, J., Majluf-Cruz, A., 2010. The C677T polymorphism of the methylenetetrahydrofolate reductase gene is associated with idiopathic ischemic stroke in the young Mexican-Mestizo population. Cerebrovasc. Dis. 29 (5), 454–459. doi: 10.1159/000289349.

Jimenez-Castro, L., Raventos-Vorst, H., Escamilla, M., 2011. Substance use disorder and schizophrenia: prevalence and sociodemographic characteristics in the Latin American population. Actas Esp Psiquiatr. 39 (2), 123–130.

Junqueira, R., Cordeiro, Q., Meira-Lima, I., Gattaz, W.F., Vallada, H., 2004. Allelic association analysis of phospholipase A2 genes with schizophrenia. Psychiatr. Genet. 14 (3), 157–160.

Karam, S.M., Barros, A.J., Matijasevich, A., Dos Santos, I.S., Anselmi, L., Barros, F., Black, M.M., 2016. Intellectual disability in a birth cohort: prevalence, etiology, and determinants at the age of 4 years. Public Health Genomics 19 (5), 290–297. doi: 10.1159/000448912.

Kisely, S., Alichniewicz, K.K., Black, E.B., Siskind, D., Spurling, G., Toombs, M., 2017. The prevalence of depression and anxiety disorders in indigenous people of the Americas: a systematic review and meta-analysis. J. Psychiatr. Res. 84, 137–152. doi: 10.1016/j.jpsychires.2016.09.032.

Kohlrausch, F.B., Gama, C.S., Lobato, M.I., Belmonte-de-Abreu, P., Callegari-Jacques, S.M., Gesteira, A., Hutz, M.H., 2008a. Naturalistic pharmacogenetic study of treatment resistance to typical neuroleptics in European-Brazilian schizophrenics. Pharmacogenet. Genomics 18 (7), 599–609. doi: 10.1097/FPC.0b013e328301a763.

Kohlrausch, F.B., Giori, I.G., Melo-Felippe, F.B., Vieira-Fonseca, T., Velarde, L.G., de Salles Andrade, J.B., Fontenelle, L.F., 2016. Association of GRIN2B gene polymorphism and obsessive compulsive disorder and symptom dimensions: a pilot study. Psychiatry Res. 243, 152–155. doi: 10.1016/j.psychres.2016.06.027.

Kohlrausch, F.B., Salatino-Oliveira, A., Gama, C.S., Lobato, M.I., Belmonte-de-Abreu, P., Hutz, M.H., 2008b. G-protein gene 825C > T polymorphism is associated with response to clozapine in Brazilian schizophrenics. Pharmacogenomics 9 (10), 1429–1436. doi: 10.2217/14622416.9.10.1429.

Kohlrausch, F.B., Salatino-Oliveira, A., Gama, C.S., Lobato, M.I., Belmonte-de-Abreu, P., Hutz, M.H., 2010. Influence of serotonin transporter gene polymorphisms on clozapine response in Brazilian schizophrenics. J. Psychiatr. Res. 44 (16), 1158–1162. doi: 10.1016/j.jpsychires.2010.04.003.

Kohlrausch, F.B., Severino-Gama, C., Lobato, M.I., Belmonte-de-Abreu, P., Carracedo, A., Hutz, M.H., 2013. The CYP1A2-163C > A polymorphism is associated with clozapine-induced generalized tonic-clonic seizures in Brazilian schizophrenia patients. Psychiatry Res. 209 (2), 242–245. doi: 10.1016/j.psychres.2013.02.030.

Kohn, R., Levav, I., de Almeida, J.M., Vicente, B., Andrade, L., Caraveo-Anduaga, J.J., Saraceno, B., 2005. Mental disorders in Latin America and the Caribbean: a public health priority. Rev. Panam Salud Publica 18 (4-5), 229–240.

Kolar, D.R., Rodriguez, D.L., Chams, M.M., Hoek, H.W., 2016. Epidemiology of eating disorders in Latin America: a systematic review and meta-analysis. Curr. Opin. Psychiatry 29 (6), 363–371, 10.1097/YCO.0000000000000279.

Kortmann, G.L., Contini, V., Bertuzzi, G.P., Mota, N.R., Rovaris, D.L., Paixao-Cortes, V.R., Bau, C.H., 2013. The role of a mineralocorticoid receptor gene functional polymorphism in the symptom dimensions of persistent ADHD. Eur. Arch. Psychiatry Clin. Neurosci. 263 (3), 181–188. doi: 10.1007/s00406-012-0321-z.

Kremeyer, B., Garcia, J., Muller, H., Burley, M.W., Herzberg, I., Parra, M.V., Ruiz-Linares, A., 2010. Genome-wide linkage scan of bipolar disorder in a Colombian population isolate replicates Loci on chromosomes 7p21-22, 1p31, 16p12 and 21q21-22 and identifies a novel locus on chromosome 12q. Hum. Hered. 70 (4), 255–268, 10.1159/000320914.

Lalli, M.A., Cox, H.C., Arcila, M.L., Cadavid, L., Moreno, S., Garcia, G., Lopera, F., 2014. Origin of the PSEN1 E280A mutation causing early-onset Alzheimer's disease. Alzheimers Dement 10 (Suppl. 5), S277–S283.e210. doi: 10.1016/j.jalz.2013.09.005.

Lek, M., Karczewski, K.J., Minikel, E.V., Samocha, K.E., Banks, E., Fennell, T., 2016. Exome Aggregation Consortium Analysis of protein-coding genetic variation in 60,706 humans. Nature 536 (7616), 285–291. doi: 10.1038/nature19057.

Li, J.Z., Absher, D.M., Tang, H., Southwick, A.M., Casto, A.M., Ramachandran, S., Myers, R.M., 2008. Worldwide human relationships inferred from genome-wide patterns of variation. Science 319 (5866), 1100–1104. doi: 10.1126/science.1153717.

Lima, I.M., Barros, A., Rosa, D.V., Albuquerque, M., Malloy-Diniz, L., Neves, F.S., de Miranda, D.M., 2015. Analysis of telomere attrition in bipolar disorder. J. Affect Disord. 172, 43–47. doi: 10.1016/j.jad.2014.09.043.

Longo, D., Schuler-Faccini, L., Brandalize, A.P., dos Santos Riesgo, R., Bau, C.H., 2009. Influence of the 5-HTTLPR polymorphism and environmental risk factors in a Brazilian sample of patients with autism spectrum disorders. Brain Res. 1267, 9–17. doi: 10.1016/j.brainres.2009.02.072.

Lopez-Narvaez, M.L., Tovilla-Zarate, C.A., González-castro, T.B., Juarez-Rojop, I., Pool-Garcia, S., Genis, A., Fresan, A., 2015. Association analysis of TPH-1 and TPH-2 genes with suicidal behavior in patients with attempted suicide in Mexican population. Compr. Psychiatry 61, 72–77. doi: 10.1016/j.comppsych.2015.05.002.

Magno, L.A., Miranda, D.M., Neves, F.S., Pimenta, G.J., Mello, M.P., De Marco, L.A., Romano-Silva, M.A., 2010. Association between AKT1 but not AKTIP genetic variants and increased risk for suicidal behavior in bipolar patients. Genes. Brain Behav. 9 (4), 411–418. doi: 10.1111/j.1601-183X.2010.00571.x.

Marquez, L., Camarena, B., Hernandez, S., Loyzaga, C., Vargas, L., Nicolini, H., 2013. Association study between BDNF gene variants and Mexican patients with obsessive-compulsive disorder. Eur. Neuropsychopharmacol. 23 (11), 1600–1605. doi: 10.1016/j.euroneuro.2013.08.001.

Martínez-levy, G.A., Benjet, C., Perez-Molina, A., Gomez-Sanchez, A., Briones-Velasco, M., Cardenas-Godinez, M., Cruz-Fuentes, C.S., 2013. Association of the catechol-O-methyltransferase gene and attention deficit hyperactivity disorder: results from an epidemiological study of adolescents of Mexico City. Psychiatr. Genet. 23 (2), 90–91. doi: 10.1097/YPG.0b013e32835d7099.

Medina, M.T., Suzuki, T., Alonso, M.E., Duron, R.M., Martinez-Juarez, I.E., Bailey, J.N., Delgado-Escueta, A.V., 2008. Novel mutations in Myoclonin1/EFHC1 in sporadic and familial juvenile myoclonic epilepsy. Neurology 70 (22 Pt 2), 2137–2144, 10.1212/01.wnl.0000313149.73035.99.

Meira-Lima, I., Michelon, L., Cordeiro, Q., Cho, H.J., Vallada, H., 2005. Allelic association analysis of the functional insertion/deletion polymorphism in the promoter region of the serotonin transporter gene in bipolar affective disorder. J. Mol. Neurosci. 27 (2), 219–224. doi: 10.1385/jmn:27:2:219.

Meira-Lima, I., Shavitt, R.G., Miguita, K., Ikenaga, E., Miguel, E.C., Vallada, H., 2004. Association analysis of the catechol-o-methyltransferase (COMT), serotonin transporter (5-HTT) and serotonin 2A receptor (5HT2A) gene polymorphisms with obsessive-compulsive disorder. Genes Brain Behav. 3 (2), 75–79.

Michelon, L., Meira-Lima, I., Cordeiro, Q., Miguita, K., Breen, G., Collier, D., Vallada, H., 2006. Association study of the INPP1, 5HTT, BDNF, AP-2beta and GSK-3beta GENE variants and restrospectively scored response to lithium prophylaxis in bipolar disorder. Neurosci. Lett. 403 (3), 288–293. doi: 10.1016/j.neulet.2006.05.001.

Miguita, K., Cordeiro, Q., Shavitt, R.G., Miguel, E.C., Vallada, H., 2006. Association study between the 1287 A/G exonic polymorphism of the norepinephrine transporter (NET) gene and obsessive-compulsive disorder in a Brazilian sample. Rev. Bras. Psiquiatr 28 (2), 158–159, S1516-44462006000200017.

Miguita, K., Cordeiro, Q., Siqueira-Roberto, J., Shavitt, R.G., Castillo, J.C., Castillo, A.R., Vallada, H., 2007. Association analysis between a VNTR intron 8 polymorphism of the dopamine transporter gene (SLC6A3) and obsessive- compulsive disorder in a Brazilian sample. Arq Neuropsiquiatr. 65 (4A), 936–941.

Mitchell, C., Hobcraft, J., McLanahan, S.S., Siegel, S.R., Berg, A., Brooks-Gunn, J., Notterman, D., 2014. Social disadvantage, genetic sensitivity, and children's telomere length. Proc. Natl. Acad. Sci. U.S.A. 111 (16), 5944–5949. doi: 10.1073/pnas.1404293111.

Mitropoulos, K., Al Jaibeji, H., Forero, D.A., Laissue, P., Wonkam, A., Lopez-Correa, C., Patrinos, G.P., 2015. Success stories in genomic medicine from resource-limited countries. Hum. Genomics 9, 11. doi: 10.1186/s40246-015-0033-3.

Morales-Marin, M.E., Genis-Mendoza, A.D., Tovilla-Zarate, C.A., Lanzagorta, N., Escamilla, M., Nicolini, H., 2016. Association between obesity and the brain-derived neurotrophic factor gene polymorphism Val66Met in individuals with bipolar disorder in Mexican population. Neuropsychiatr. Dis. Treat. 12, 1843–1848. doi: 10.2147/NDT.S104654.

Moreira, F.P., Fabiao, J.D., Bittencourt, G., Wiener, C.D., Jansen, K., Oses, J.P., Ghisleni, G., 2015. The Met allele of BDNF Val66Met polymorphism is associated with increased BDNF levels in generalized anxiety disorder. Psychiatr. Genet. 25 (5), 201–207. doi: 10.1097/ypg.0000000000000097.

Moreno-Estrada, A., Gignoux, C.R., Fernandez-Lopez, J.C., Zakharia, F., Sikora, M., Contreras, A.V., Bustamante, C.D., 2014. Human genetics. The genetics of Mexico recapitulates Native

American substructure and affects biomedical traits. Science 344 (6189), 1280–1285. doi: 10.1126/science.1251688.

Mota, N.R., Rovaris, D.L., Bertuzzi, G.P., Contini, V., Vitola, E.S., Grevet, E.H., Bau, C.H., 2013. DRD2/DRD4 heteromerization may influence genetic susceptibility to alcohol dependence. Mol. Psychiatry 18 (4), 401–402. doi: 10.1038/mp.2012.50.

Mota, N.R., Rovaris, D.L., Kappel, D.B., Picon, F.A., Vitola, E.S., Salgado, C.A., Bau, C.H., 2015. NCAM1-TTC12-ANKK1-DRD2 gene cluster and the clinical and genetic heterogeneity of adults with ADHD. Am. J. Med. Genet. B Neuropsychiatr. Genet., 10.1002/ajmg.b.32317.

Moura, R.R., Coelho, A.V., Balbino Vde, Q., Crovella, S., Brandao, L.A., 2015. Meta-analysis of Brazilian genetic admixture and comparison with other Latin America countries. Am. J. Hum. Biol. 27 (5), 674–680. doi: 10.1002/ajhb.22714.

Muchnik, C., Olivar, N., Dalmasso, M.C., Azurmendi, P.J., Liberczuk, C., Morelli, L., Brusco, L.I., 2015. Identification of PSEN2 mutation p.N141I in Argentine pedigrees with early-onset familial Alzheimer's disease. Neurobiol. Aging 36 (10)doi: 10.1016/j.neurobiolaging.2015.06.011, 2674-2677.e2671.

Murray, C.J., Vos, T., Lozano, R., Naghavi, M., Flaxman, A.D., Michaud, C., Memish, Z.A., 2012. Disability-adjusted life years (DALYs) for 291 diseases and injuries in 21 regions, 1990–2010: a systematic analysis for the Global Burden of Disease Study 2010. Lancet 380 (9859), 2197–2223. doi: 10.1016/S0140-6736(12)61689-4.

Nag, A., Bochukova, E.G., Kremeyer, B., Campbell, D.D., Muller, H., Valencia-Duarte, A.V., Ruiz-Linares, A., 2013. CNV analysis in Tourette syndrome implicates large genomic rearrangements in COL8A1 and NRXN1. PLoS One 8 (3), e59061. doi: 10.1371/journal.pone.0059061.

Neves, F.S., Malloy-Diniz, L., Romano-Silva, M.A., Campos, S.B., Miranda, D.M., De Marco, L., Correa, H., 2011. The role of BDNF genetic polymorphisms in bipolar disorder with psychiatric comorbidities. J. Affect. Disord. 131 (1–3), 307–311. doi: 10.1016/j.jad.2010.11.022.

Nitrini, R., Bottino, C.M., Albala, C., Custodio Capunay, N.S., Ketzoian, C., Llibre Rodriguez, J.J., Caramelli, P., 2009. Prevalence of dementia in Latin America: a collaborative study of population-based cohorts. Int. Psychogeriatr. 21 (4), 622–630. doi: 10.1017/S1041610209009430.

Ojeda, D.A., Nino, C.L., Lopez-Leon, S., Camargo, A., Adan, A., Forero, D.A., 2014a. A functional polymorphism in the promoter region of MAOA gene is associated with daytime sleepiness in healthy subjects. J. Neurol. Sci. 337 (1–2), 176–179. doi: 10.1016/j.jns.2013.12.005.

Ojeda, D.A., Perea, C.S., Nino, C.L., Gutierrez, R.M., Lopez-Leon, S., Arboleda, H., Forero, D.A., 2013. A novel association of two non-synonymous polymorphisms in PER2 and PER3 genes with specific diurnal preference subscales. Neurosci. Lett. 553, 52–56. doi: 10.1016/j.neulet.2013.08.016.

Ojeda, D.A., Perea, C.S., Suarez, A., Nino, C.L., Gutierrez, R.M., Lopez-Leon, S., Forero, D.A., 2014b. Common functional polymorphisms in SLC6A4 and COMT genes are associated with circadian phenotypes in a South American sample. Neurol. Sci. 35 (1), 41–47. doi: 10.1007/s10072-013-1466-x.

Ortega-Rojas, J., Morales, L., Guerrero, E., Arboleda-Bustos, C.E., Mejia, A., Forero, D., Lopez, L., Pardo, R., Arboleda, G., Yunis, J., Arboleda, H., 2016. Association Analysis of Polymorphisms in TOMM40, CR1, PVRL2, SORL1, PICALM, and 14q32.13 Regions in Colombian Alzheimer Disease Patients. Alzheimer Dis. Assoc. Disord. 30 (4), 305–309. doi: 10.1097/WAD.0000000000000142.

Ota, V.K., Spindola, L.N., Gadelha, A., dos Santos Filho, A.F., Santoro, M.L., Christofolini, D.M., Belangero, S.I., 2012. DRD1 rs4532 polymorphism: a potential pharmacogenomic marker for treatment response to antipsychotic drugs. Schizophr. Res. 142 (1–3), 206–208. doi: 10.1016/j.schres.2012.08.003.

Parra, F.C., Amado, R.C., Lambertucci, J.R., Rocha, J., Antunes, C.M., Pena, S.D., 2003. Color and genomic ancestry in Brazilians. Proc. Natl. Acad. Sci. U.S.A. 100 (1), 177–182. doi: 10.1073/pnas.0126614100.

Pena, S.D., Di Pietro, G., Fuchshuber-Moraes, M., Genro, J.P., Hutz, M.H., Kehdy Fde, S., Suarez-Kurtz, G., 2011. The genomic ancestry of individuals from different geographical regions of Brazil is more uniform than expected. PLoS One 6 (2), e17063. doi: 10.1371/journal. pone.0017063.

Peñas-lledó, E.M., Trejo, H.D., Dorado, P., Ortega, A., Jung, H., Alonso, E., Llerena, A., 2013. CYP2D6 ultrarapid metabolism and early dropout from fluoxetine or amitriptyline monotherapy treatment in major depressive patients. Mol. Psychiatry 18 (1), 8–9. doi: 10.1038/ mp.2012.91.

Peralta-Leal, V., Leal-Ugarte, E., Meza-Espinoza, J.P., Gutierrez-Angulo, M., Hernandez-Benitez, C.T., Garcia-Rodriguez, A., Duran-Gonzalez, J., 2012. Association of serotonin transporter gene polymorphism 5-HTTLPR and depressive disorder in a Mexican population. Psychiatr. Genet. 22 (5), 265–266. doi: 10.1097/YPG.0b013e32834f3577.

Perea, C.S., Nino, C.L., Lopez-Leon, S., Gutierrez, R., Ojeda, D., Arboleda, H., Forero, D.A., 2014. Study of a functional polymorphism in the PER3 gene and diurnal preference in a Colombian Sample. Open Neurol. J. 8, 7–10. doi: 10.2174/1874205x01408010007.

Pereira, P.A., Bicalho, M.A., de Moraes, E.N., Malloy-Diniz, L., Bozzi, I.C., Nicolato, R., Romano-Silva, M.A., 2014. Genetic variant of AKT1 and AKTIP associated with late-onset depression in a Brazilian population. Int. J. Geriatr. Psychiatry 29 (4), 399–405. doi: 10.1002/gps.4018.

Pereira, P.A., Romano-Silva, M.A., Bicalho, M.A., de Moraes, E.N., Malloy-Diniz, L., Pimenta, G.J., Miranda, D.M., 2012. Catechol-O-methyltransferase genetic variant associated with the risk of Alzheimer's disease in a Brazilian population. Dement Geriatr. Cogn. Disord. 34 (2), 90–95. doi: 10.1159/000341578.

Pereira Pde, A., Romano-Silva, M.A., Bicalho, M.A., De Marco, L., Correa, H., de Campos, S.B., de Miranda, D.M., 2011. Association between tryptophan hydroxylase-2 gene and late-onset depression. Am. J. Geriatr. Psychiatry 19 (9), 825–829. doi: 10.1097/JGP.0b013e31820eeb21.

Polanczyk, G., Zeni, C., Genro, J.P., Guimaraes, A.P., Roman, T., Hutz, M.H., Rohde, L.A., 2007. Association of the adrenergic alpha2A receptor gene with methylphenidate improvement of inattentive symptoms in children and adolescents with attention-deficit/hyperactivity disorder. Arch. Gen. Psychiatry 64 (2), 218–224. doi: 10.1001/archpsyc.64.2.218.

Prestes, A.P., Marques, F.Z., Hutz, M.H., Bau, C.H., 2007. The GNB3 C825T polymorphism and depression among subjects with alcohol dependence. J. Neural Transm. (Vienna) 114 (4), 469–472. doi: 10.1007/s00702-006-0550-2.

Prince, M., Bryce, R., Albanese, E., Wimo, A., Ribeiro, W., Ferri, C.P., 2013. The global prevalence of dementia: a systematic review and metaanalysis. Alzheimers Dement 9 (1), 63–75.e62. doi: 10.1016/j.jalz.2012.11.007.

Pringsheim, T., Jette, N., Frolkis, A., Steeves, T.D., 2014. The prevalence of Parkinson's disease: a systematic review and meta-analysis. Mov Disord. 29 (13), 1583–1590. doi: 10.1002/mds.25945.

Quinones, L.A., Lavanderos, M.A., Cayun, J.P., Garcia-Martin, E., Agundez, J.A., Caceres, D.D., Lares-Assef, I., 2014. Perception of the usefulness of drug/gene pairs and barriers for pharmacogenomics in Latin America. Curr. Drug Metab. 15 (2), 202–208.

Reynolds, G.P., Fachim, H.A., 2016. Does DNA methylation influence the effects of psychiatric drugs? Epigenomics 8 (3), 309–312. doi: 10.2217/epi.15.116.

Rocha, F.F., Alvarenga, N.B., Lage, N.V., Romano-Silva, M.A., Marco, L.A., Correa, H., 2011. Associations between polymorphic variants of the tryptophan hydroxylase 2 gene and obsessive-compulsive disorder. Rev. Bras. Psiquiatr. 33 (2), 176–180.

Rocha, F.F., Malloy-Diniz, L., Lage, N.V., Correa, H., 2010. Positive association between MET allele (BDNF Val66Met polymorphism) and obsessive-compulsive disorder. Rev. Bras. Psiquiatr. 32 (3), 323–324.

Rocha, F.F., Marco, L.A., Romano-Silva, M.A., Corrêa, H., 2009. Obsessive-compulsive disorder and 5-HTTLPR. Rev. Bras. Psiquiatr. 31 (3), 287–288, PMID: 19784500.

Rodeiro, I., Remirez-Figueredo, D., Garcia-Mesa, M., Dorado, P., Lerena, A.L., Pharmacogenetics CEIBA.FP Consortium of the Ibero-American Network of Pharmacogenetics and Pharmacogenomics RIBEF, 2012. Pharmacogenetics in Latin American populations: regulatory aspects, application to herbal medicine, cardiovascular and psychiatric disorders. Drug Metab. Drug Interact. 27 (1), 57–60. doi: 10.1515/dmdi-2012-0006.

Roman, T., Schmitz, M., Polanczyk, G.V., Eizirik, M., Rohde, L.A., Hutz, M.H., 2003. Is the alpha-2A adrenergic receptor gene (ADRA2A) associated with attention-deficit/hyperactivity disorder? Am. J. Med. Genet. B Neuropsychiatr. Genet. 120B (1), 116–120. doi: 10.1002/ajmg.b.20018.

Ruiz-Linares, A., Adhikari, K., Acuna-Alonzo, V., Quinto-Sanchez, M., Jaramillo, C., Arias, W., Gonzalez-Jose, R., 2014. Admixture in Latin America: geographic structure, phenotypic diversity and self-perception of ancestry based on 7,342 individuals. PLoS Genet. 10 (9), e1004572. doi: 10.1371/journal.pgen.1004572.

Salatino-Oliveira, A., Akutagava-Martins, G.C., Bruxel, E.M., Genro, J.P., Polanczyk, G.V., Zeni, C., Hutz, M.H., 2016a. NOS1 and SNAP25 polymorphisms are associated with Attention-Deficit/Hyperactivity Disorder symptoms in adults but not in children. J. Psychiatr. Res. 75, 75–81. doi: 10.1016/j.jpsychires.2016.01.010.

Salatino-Oliveira, A., Genro, J.P., Chazan, R., Zeni, C., Schmitz, M., Polanczyk, G., Hutz, M.H., 2012a. Association study of GIT1 gene with attention-deficit hyperactivity disorder in Brazilian children and adolescents. Genes Brain Behav. 11 (7), 864–868. doi: 10.1111/j.1601-183X.2012.00835.x.

Salatino-Oliveira, A., Genro, J.P., Guimaraes, A.P., Chazan, R., Zeni, C., Schmitz, M., Hutz, M.H., 2012b. Cathechol-O-methyltransferase Val(158)Met polymorphism is associated with disruptive behavior disorders among children and adolescents with ADHD. J. Neural Transm. (Vienna) 119 (6), 729–733. doi: 10.1007/s00702-012-0766-2.

Salatino-Oliveira, A., Genro, J.P., Polanczyk, G., Zeni, C., Schmitz, M., Kieling, C., Hutz, M.H., 2015. Cadherin-13 gene is associated with hyperactive/impulsive symptoms in attention/deficit hyperactivity disorder. Am. J. Med. Genet. B Neuropsychiatr. Genet. 168B (3), 162–169. doi: 10.1002/ajmg.b.32293.

Salatino-Oliveira, A., Genro, J.P., Zeni, C., Polanczyk, G.V., Chazan, R., Guimaraes, A.P., Hutz, M.H., 2011. Catechol-O-methyltransferase valine158methionine polymorphism moderates methylphenidate effects on oppositional symptoms in boys with attention-deficit/hyperactivity disorder. Biol. Psychiatry 70 (3), 216–221. doi: 10.1016/j.biopsych.2011.03.025.

Salatino-Oliveira, A., Wagner, F., Akutagava-Martins, G.C., Bruxel, E.M., Genro, J.P., Zeni, C., Hutz, M.H., 2016b. MAP1B and NOS1 genes are associated with working memory in youths with attention-deficit/hyperactivity disorder. Eur. Arch. Psychiatry Clin. Neurosci. 266 (4), 359–366. doi: 10.1007/s00406-015-0626-9.

Santos, I.S., Barros, A.J., Matijasevich, A., Zanini, R., Chrestani Cesar, M.A., Camargo-Figuera, F.A., Victora, C.G., 2014. Cohort profile update: 2004 Pelotas (Brazil) Birth Cohort Study. Body composition, mental health and genetic assessment at the 6 years follow-up. Int. J. Epidemiol. 43 (5), 1437–11437. doi: 10.1093/ije/dyu144.

Scharf, J.M., Yu, D., Mathews, C.A., Neale, B.M., Stewart, S.E., Fagerness, J.A., Pauls, D.L., 2013. Genome-wide association study of Tourette's syndrome. Mol. Psychiatry 18 (6), 721–728. doi: 10.1038/mp.2012.69.

Segal, J., Schenkel, L.C., Oliveira, M.H., Salum, G.A., Bau, C.H., Manfro, G.G., Leistner-Segal, S., 2009. Novel allelic variants in the human serotonin transporter gene linked polymorphism (5-HTTLPR) among depressed patients with suicide attempt. Neurosci. Lett. 451 (1), 79–82. doi: 10.1016/j.neulet.2008.12.015.

Sepulveda-Falla, D., Glatzel, M., Lopera, F., 2012. Phenotypic profile of early-onset familial Alzheimer's disease caused by presenilin-1 E280A mutation. J. Alzheimers Dis. 32 (1), 1–12. doi: 10.3233/jad-2012-120907.

Sesarini, C.V., Costa, L., Granana, N., Coto, M.G., Pallia, R.C., Argibay, P.F., 2015. Association between GABA(A) receptor subunit polymorphisms and autism spectrum disorder (ASD). Psychiatry Res. 229 (1–2), 580–582. doi: 10.1016/j.psychres.2015.07.077.

Sesarini, C.V., Costa, L., Naymark, M., Granana, N., Cajal, A.R., Garcia Coto, M., Argibay, P.F., 2014. Evidence for interaction between markers in GABA(A) receptor subunit genes in an Argentinean autism spectrum disorder population. Autism Res. 7 (1), 162–166. doi: 10.1002/aur.1353.

Sharan, P., Gallo, C., Gureje, O., Lamberte, E., Mari, J.J., Mazzotti, G., World Health Organization-Global Forum for Health Research – Mental Health Research Mapping Project, Group, 2009. Mental health research priorities in low- and middle-income countries of Africa, Asia, Latin America and the Caribbean. Br. J. Psychiatry 195 (4), 354–363. doi: 10.1192/bjpbp.108.050187.

Solís-ortiz, S., Perez-Luque, E., Morado-Crespo, L., Gutierrez-Munoz, M., 2010. Executive functions and selective attention are favored in middle-aged healthy women carriers of the Val/Val genotype of the catechol-o-methyltransferase gene: a behavioral genetic study. Behav. Brain Funct. 6, 67. doi: 10.1186/1744-9081-6-67.

Sosa-Macias, M., Teran, E., Waters, W., Fors, M.M., Altamirano, C., Jung-Cook, H.A., Lerena, L., 2016. Pharmacogenetics and ethnicity: relevance for clinical implementation, clinical trials, pharmacovigilance and drug regulation in Latin America. Pharmacogenomics, 10.2217/pgs-2016-0153.

Speck-Hernandez, C.A., Ojeda, D.A., Castro-Vega, L.J., Forero, D.A., 2015. Relative telomere length is associated with a functional polymorphism in the monoamine oxidase A gene in a South American sample. J. Genet. 94 (2), 305–308.

Stewart, S.E., Yu, D., Scharf, J.M., Neale, B.M., Fagerness, J.A., Mathews, C.A., Pauls, D.L., 2013. Genome-wide association study of obsessive-compulsive disorder. Mol. Psychiatry 18 (7), 788–798. doi: 10.1038/mp.2012.85.

Suarez-Kurtz, G., 2004. Pharmacogenomics in admixed populations: the Brazilian pharmacogenetics/pharmacogenomics network--REFARGEN. Pharmacogenomics J. 4 (6), 347–348. doi: 10.1038/sj.tpj.6500287.

Suarez-Kurtz, G., Pena, S.D., 2006. Pharmacogenomics in the Americas: the impact of genetic admixture. Curr. Drug Targets 7 (12), 1649–1658.

Sudmant, P.H., Rausch, T., Gardner, E.J., Handsaker, R.E., Abyzov, A., Huddleston, J., Korbel, J.O., 2015. An integrated map of structural variation in 2,504 human genomes. Nature 526 (7571), 75–81. doi: 10.1038/nature15394.

Tovilla-Zarate, C.A., Vargas, I., Hernandez, S., Fresan, A., Aguilar, A., Escamilla, R., Camarena, B., 2014. Association study between the MDR1 gene and clinical characteristics in schizophrenia. Rev. Bras. Psiquiatr. 36 (3), 227–232.

Tovilla-Zarate, C., Medellin, B.C., Fresan, A., Lopez-Narvaez, L., Castro, T.B., Juarez Rojop, I., Nicolini, H., 2013. No association between catechol-o-methyltransferase Val108/158Met polymorphism and schizophrenia or its clinical symptomatology in a Mexican population. Mol. Biol. Rep. 40 (2), 2053–2058. doi: 10.1007/s11033-012-2264-x.

Tovo-Rodrigues, L., Rohde, L.A., Menezes, A.M., Polanczyk, G.V., Kieling, C., Genro, J.P., Hutz, M.H., 2013. DRD4 rare variants in attention-deficit/hyperactivity disorder (ADHD): further evidence from a birth cohort study. PLoS One 8 (12), e85164. doi: 10.1371/journal.pone.0085164.

Tovo-Rodrigues, L., Rohde, L.A., Roman, T., Schmitz, M., Polanczyk, G., Zeni, C., Hutz, M.H., 2012. Is there a role for rare variants in DRD4 gene in the susceptibility for ADHD? Searching for an effect of allelic heterogeneity. Mol. Psychiatry 17 (5), 520–526. doi: 10.1038/mp.2011.12.

Urraca, N., Camarena, B., Aguilar, A., Fresan, A., Apiquian, R., Orozco, L., Nicolini, H., 2011. Association study of DRD3 gene in schizophrenia in Mexican sib-pairs. Psychiatry Res. 190 (2–3), 367–368. doi: 10.1016/j.psychres.2011.06.009.

Vasconcelos, A.C., Neto Ede, S., Pinto, G.R., Yoshioka, F.K., Motta, F.J., Vasconcelos, D.F., Canalle, R., 2015. Association study of the SLC6A3 VNTR (DAT) and DRD2/ANKK1 Taq1A polymorphisms with alcohol dependence in a population from northeastern Brazil. Alcohol Clin. Exp. Res. 39 (2), 205–211. doi: 10.1111/acer.12625.

Vieira, R.N., Avila, R., de Paula, J.J., Cintra, M.T., de Souza, R.P., Nicolato, R., Bicalho, M.A., 2016. Association between DCHS2 gene and mild cognitive impairment and Alzheimer's disease in an elderly Brazilian sample. Int. J. Geriatr. Psychiatry 31 (12), 1337–1344. doi: 10.1002/gps.4440.

Vieira, R.N., Magalhães, J.D., Sant'Anna, J., Moriguti, M.M., de Paula, J.J., Cintra, M.T., de Miranda, D.M., De Marco, L., de Moraes, E.N., Romano-Silva, M.A., Bicalho, MA., 2015. The GAB2 and BDNF polymorphisms and the risk for late-onset Alzheimer's disease in an elderly Brazilian sample. Int. Psychogeriatr. 27 (10), 1687–1692. doi: 10.1017/S1041610215000514.

Vigo, D., Thornicroft, G., Atun, R., 2016. Estimating the true global burden of mental illness. Lancet Psychiatry 3 (2), 171–178. doi: 10.1016/S2215-0366(15)00505-2.

Vos, T., Flaxman, A.D., Naghavi, M., Lozano, R., Michaud, C., Ezzati, M., Memish, Z.A., 2012. Years lived with disability (YLDs) for 1160 sequelae of 289 diseases and injuries 1990-2010: a systematic analysis for the Global Burden of Disease Study 2010. Lancet 380 (9859), 2163–2196. doi: 10.1016/S0140-6736(12)61729-2.

Wang, S., Ray, N., Rojas, W., Parra, M.V., Bedoya, G., Gallo, C., Ruiz-Linares, A., 2008. Geographic patterns of genome admixture in Latin American Mestizos. PLoS Genet 4 (3), e1000037. doi: 10.1371/journal.pgen.1000037.

Whiteford, H.A., Ferrari, A.J., Degenhardt, L., Feigin, V., Vos, T., 2015. The global burden of mental, neurological and substance use disorders: an analysis from the Global Burden of Disease Study 2010. PLoS One 10 (2), e0116820. doi: 10.1371/journal.pone.0116820.

Zeni, C.P., Guimaraes, A.P., Polanczyk, G.V., Genro, J.P., Roman, T., Hutz, M.H., Rohde, L.A., 2007. No significant association between response to methylphenidate and genes of the dopaminergic and serotonergic systems in a sample of Brazilian children with attention-deficit/hyperactivity disorder. Am. J. Med. Genet. B Neuropsychiatr. Genet. 144B (3), 391–394. doi: 10.1002/ajmg.b.30474.

Zeni, C.P., Tramontina, S., Aguiar, B.W., Salatino-Oliveira, A., Pheula, G.F., Sharma, A., Rohde, L.A., 2016. BDNF Val66Met polymorphism and peripheral protein levels in pediatric bipolar disorder and attention-deficit/hyperactivity disorder. Acta Psychiatr. Scand. 134 (3), 268–274. doi: 10.1111/acps.12587.

Zoghbi, H.Y., Warren, S.T., 2010. Neurogenetics: advancing the "next-generation" of brain research. Neuron 68 (2), 165–173. doi: 10.1016/j.neuron.2010.10.015.

# CHAPTER 7

# Experience in the Development of Genomics Companies During the Last 20 Years in Argentina

**Viviana Ada Bernath\*, Mariana Herrera Piñero\*\***

*\*Laboratorio Genda sa, Buenos Aires, Argentina \*\*Banco Nacional de Datos Genéticos, Buenos Aires, Argentina*

A couple of months ago we received a call from the city of Ushuaia. It was Camila telling us that her sister Mariela, 33, had suddenly died. Mariela was the mother of a 3-month-old baby girl and a 4-year-old boy. The whole family was looking for an explanation of her unexpected death. In particular, the two children's father was really worried about the possibility that the children would have inherited some illness from their mother. To carry out the genetic testing, we needed a sample of Mariela's blood. Fortunately for everyone, we were able to obtain the sample, and the results were ready a month later. We found a mutation[1] in Mariela's DNA[2] compatible with an arrhythmia associated with many cases of sudden death in postpartum women. The question was: How do we continue? Should we study the children or not? The family asked us to do so, and we detected that both children had inherited their mother's mutation. However, the good news was that having such information so early could help them to care for themselves and prevent a sudden arrhythmia death in the future. In fact, cardiologists are already considering the possibility of implanting automatic defibrillators and prescribing beta-blocker treatments. In conclusion, if these kids follow the medical recommendations, they and their family will be able to live without worry.

Thinking about solving cases like Mariela's 15, 10, or even 5 years ago would have been a matter of "science fiction." However, today, as never before, genomics have opened up a world of solutions that is changing the lives of millions of people.

---

[1] A sudden departure from the parent type in one or more heritable characteristics, caused by a change in a gene or a chromosome.
[2] Deoxyribonucleic acid, an extremely long macromolecule that is the main component of chromosomes and is the material that transfers genetic characteristics in all life forms.

**131**

Genomic Medicine in Emerging Economies. http://dx.doi.org/10.1016/B978-0-12-811531-2.00007-2

In this chapter, and by relating our own experiences as PhDs in molecular biology—specializing in human genetics and forensic genetics—we will try to synthesize how genomics advances of the last 30 years have been implemented in Argentina.

## A BIT OF HISTORY

Less than 100 years had passed between 1869, when the Swiss chemist Friedrich Miescher isolated a milky substance from the white blood cells present in the purulent secretions of discarded bandages after wounded soldiers were healed (which he called "nuclein"), and 1953, when Watson and Crick described the structure of DNA (https://www.ndsu.edu/pubweb/~mcclean/plsc411/History-of-Genetics-and-Genomics-narrative-and-overheads.pdf). During that period of time, Walter Flemming, a German physiologist who was dedicated to the study of cellular structures, discovered that the cell nucleus contains a substance that is tinged with color, which he called "chromatin;" this substance was able to separate during cell division into filaments, which received the name of "chromosomes." Thomas Hunt Morgan, while working with the fruit fly, confirmed the Mendelian hypothesis that genes are located on chromosomes; while Avery, MacLeod, and McCarty, during their work with bacteria, were able to prove that DNA was the vehicle to biological specificity and that this molecule resided in human genetic information. They also claimed that DNA was the gene and chromosome genetic material, and through DNA certain characteristics were inherited from parents and transmitted to children, and so on, from generation to generation.

Since then, the concepts of inheritance, genes, DNA, and chromosomes have merged. Scientific thinking has finally managed to bring them all together into one box.

Thus in 1953 the Age of Genomics began (Smith, 2005; https://www.future-learn.com/courses/the-genomics-era/0/steps/4866).

In all societies, scientific policies are correlated in one way or another with past and present political moments. That is why in order to understand what happened with genomics in Argentina in recent years, especially in the private sector, we need a brief historical account.

From 1930 to 1983, Argentina suffered numerous coups d'état. On July 29, 1966, during the de facto government of General Ongania, the episode known as the "The Night of the Long Police Batons" took place (Morero et al., 2002). On that day, hundreds of teachers, students, and nonteachers who occupied several of Buenos Aires University buildings in defense of university autonomy and academic freedom were savagely beaten by the security forces. The government ordered the intervention to national universities, their military occupation, and the academic "purification," that is to say, the expulsion of opposing professors, regardless of their academic level. The consequence of that black night for the national culture was the dismissal and resignation of 700 of the best professors of Argentina's universities, who continued their brilliant careers abroad, depriving our country of the most lucid minds of academic scientific knowledge.

After a brief democratic interregnum, from 1973 to 1976, during which it was not possible to recover what was lost, a new coup d'état, much more violent and merciless than the previous ones, subsumed the country and its entire scientific community in a long period of terror, persecution, and censorship (https://www.nytimes.com/2016/03/23/opinion/the-long-shadow-of- argentinas-dictatorship.html?_r=0).

It was not until 1983, with the definitive return to democracy, that many of those persecuted researchers and the new generations formed abroad returned to our country. With this process, the recovery of scientific-technological development of our institutes and universities began. Many of these professionals began with the first research lines in molecular biology applied to human genetics. From these research projects emerged the first PhD students in molecular genetics in our country, and the authors of this chapter are part thereof. Our thesis directors were also young, just a few years older than us, and they came back with the push of their postdocs and the support of the laboratories that they had worked for in Europe or the United States.

A few years later, toward the end of the 1980s, molecular genetics began to be used to diagnose diseases in different countries. The technologies used at that time were essentially manual, that is, they were not automated, and required an almost artisanal style of work, which also demanded obtaining large amounts of DNA. This meant that significant blood samples (from 10 to 20 ml) were required from the patient. The DNA was then digested or cut with restriction enzymes or scissors that only cut a certain section of the DNA. Subsequently, vertical or horizontal electrophoresis were performed to differentiate specific bands that correlated with the result search for the diagnosed patient (https://www.ncbi.nlm.nih.gov/probe/docs/techrflp/). In many cases, tedious transfer and hybridization methodologies (technically known as the Southern blot[3] method) were applied in which radioactive material was used (Southern, 1975). To observe our results, we had to enter an improvised darkroom somewhere in the laboratory to reveal an autoradiography impacted by radioactive signals, which had been incorporated into specific DNA sites, and whose signal indicated the final result. Healthy patients presented one band pattern and the other carriers or patients diagnosed with a disease presented a different one, and in that way the first diagnoses by molecular genetics were determined.

---

[3] A procedure for identifying and measuring the amount of a specific DNA sequence or gene in a mixed extract.

## THE POLYMERASE CHAIN REACTION TECHNIQUE

In 1984 the polymerase chain reaction (PCR[4] or DNA copier) technique discovery ushered in a technological leap in molecular biology, with unimaginable consequences, as it allowed the obtaining of millions of DNA copies from a sample fragment (Saiki et al., 1985). Extraction of large quantities of blood was no longer necessary, since this "biological copier" could amplify the genetic material even from minimal samples. In addition, this biological material was no longer confined to a blood sample. Samples could also be taken from oral swabs, bone remains, urine, nails, hair, etc. This was one of the great advances that brought molecular genetics to the service of disease diagnosis and forensic genetics.

> "Thus, in a very artisanal way and through the PCR technique use, the molecular genetics allowed us to start identifying the first mutations causing genetic diseases, such as cystic fibrosis or Duchenne's disease, some polymorphisms associated with coagulation factors, hemochromatosis[5], and associated microdeletions[6]

These methodologies were limited only to the search for the most frequent mutations present in these genes. Manual sequencing methods (reading of DNA sequence) were used, which were confined only to research laboratories. Thus molecular genetic tools began to be applied especially in the infectious disease field to diagnose patients with hepatitis C, hepatitis B, HIV, and some types of leukemia (Yang and Rothman, 2004).

Finally, in the early 1990s the first generation of Argentinian molecular biologists emerges, a group which we are part of.

Although the scientific-technological system was undergoing a recovery process, economic policies did not collaborate with subsidies to research, nor were researchers' salaries globally competitive. Thus many of our colleagues who had recently completed their studies or postgraduate work decided to emigrate again. At that time, in the field of research and development in human genetics, the first laboratories of molecular genetics applied to health service within public hospitals were created, where different hospital units began to create genetic diagnostic services, mainly directed at high-frequency pathologies in specialized services. Thus we can highlight the Pediatric Hospital Juan P. Garrahan where

---

[4] PCR, a technique for rapidly producing many copies of a fragment of DNA for diagnostic or research purposes.

[5] A rare metabolic disorder characterized by bronzed skin, cirrhosis, and severe diabetes, caused by the deposit in tissue, especially of the liver and pancreas, of hemosiderin and other pigments containing iron.

[6] Mutations created by deletion of very little fragments of DNA in the genome.

cystic fibrosis[7], fragile X syndrome[8], and spinal muscular atrophy diagnoses were developed (Chertkoff et al., 1997). Very rarely did researchers who had recently earned their PhDs, like us, choose to leave the public scientific-technological system. When we glimpsed the genomics future, we decided to create a private diagnostic center to treat genetic diseases by using molecular biology.

> "In this way, the first companies of disease molecular diagnosis were created, a field where we are pioneers and, in 1991, we founded the first molecular diagnostic laboratory in Argentina: Biología Molecular Diagnóstica, specialize in the detection of infectious, genetic and oncohematological diseases task that one of us continues to perform daily."

Since there were no commercial kits for this kind of diagnosis, we developed our own kits in a private laboratory. Along with other biologists who followed the same path, and in order to develop a diagnosis, we used scientific publications as reference; by replicating the methodologies we were able to fine-tune genetic tests. To validate the tests in Argentina, we signed agreements among companies and public hospitals for the analysis of a number of patients free of charge. After validating the results for these patients, we started to offer the service to all patients. In Argentina the reagents used were imported and very expensive.

> "Our task ranged from reading scientific publications to washing and autoclaving the plastic material that we had to recycle in order to reduce the test costs. However, in this corner of the world, we knew we could do it, and the desire to offer patients the latest technology encouraged us to carry out any necessary task."

While in the United States and Europe the first PCR machines[9] were being used to diagnose diseases, we created baths at a fixed temperature, clocked the time, and tried to reproduce the technique with what we had available. Fortunately, in a short time, both locally manufactured and imported PCR machines began to appear in Argentina at a reasonable cost. At that moment, we were able to increase the type and number of tests for our patients.

In the 1990s, molecular genetics and its applications in diagnosis and prevention were an unknown terrain for the Argentine medical community. Therefore, in order to perform the first tests in a laboratory, technological development was not enough. As there was no precedent for a company like ours in Argentina, it took a huge investment of time and money as well as in health professionals' education. This work was carried out through training courses in hospital services of different medical specialties. Moreover, population surveys

---

[7] A hereditary chronic disease of the exocrine glands, characterized by the production of viscid mucus that obstructs the pancreatic ducts and bronchi, leading to infection and fibrosis.

[8] A neurological inherited disorder inherited through the chromosome X.

[9] Machines used to photocopy parts of the DNA by PCR technique.

of mutations related to each pathology and projects subsidized by the state were used in the presentations of scientific works in medical congresses.

> "With an information folder under our arm, we visited the main neurology, hematology, or genetics centers of public hospitals or related private institutions, explained who we were, and waited until some professional received us. Luckily for us we were welcomed. We were bringing the future."

These first meetings often led to the arrangement of training seminars.

The first population survey was conducted in 1999 and was based on autosomal dominant polycistic kidney disease (Iglesias et al., 1997) and on neurodegenerative diseases (Duchenne and Becker muscular dystrophy, spinocerebellar ataxias, Charcott-Marie-Tooth disease type 1A[10], and Huntington's corea disease[11]) (Baranzini et al., 1998).

## FORENSIC GENETICS

Fabian was a man who had divorced and had two daughters. When he thought that he would never fall in love again, he met Marcela, a 22-year-old woman. After a while, Marcela invited him to dinner at her house to meet her family. When Fabian saw Marcela's mother, he immediately recognized her. She was Roxana, his 17-year-old girlfriend. Neither Fabian nor Roxana said they knew each other, but a few days later she called him to arrange a meeting. They met in a bar, and Roxana told him that when he left her, she had learned that she was pregnant. However, since she did not want to force him to be the father of a daughter they had not planned for, she has never told him. At that moment she told him that Marcela was his biological daughter. When Marcela heard the story, she wanted to corroborate it, and, together with Fabian, they showed up in our laboratory to obtain a paternity test. And while we took the samples, they confessed their love story. The result determined that there was a probability of 99.99% that Fabian was Marcela's biological father.

As we mentioned before, on March 24, 1976, the democratic government of the Argentine Republic was overthrown by another coup d'état. From that moment, a period called the "National Reorganization Process" began.

The military forces in power launched a repressive political system, the aim of which was to impose "order" through terror. It was the bloodiest stage in Argentine history, in which thousands of people, including students, workers, trade unionists, intellectuals, and professionals, were kidnapped by the armed forces, tortured in clandestine detention centers, and murdered. Most of these people remain missing. It is estimated that about 340 clandestine detention centers were scattered throughout the national territory. Among the

---

[10] A neurological inherited disorder.
[11] A neurological inherited disorder.

best known were Campo de Mayo, Escuela de Mecánica de la Armada, Vesuvio, Garage Olimpo, Pozo de Banfield, and La Perla. Many women were pregnant or had children when they were abducted. Children born in captivity in clandestine maternities or kidnapped with their parents were mostly held as spoils of war, given to families generally related to the armed forces, and their identities were forged as they were registered as their appropriators' children. Others were murdered, sold, or abandoned in emergency shelters for minors.

Family organizations, such as the Grandmothers of Plaza de Mayo, began the search of nearly 500 children stolen, and, when democracy returned in 1983, these organizations became the engine to develop the first statistical calculations in forensic genetics in our country. In 1987 the first genetic data bank was created, being the foundation stone of DNA banks of criminology and people search that emerged in 1995 (http://en.mincyt.gob.ar/ministerio/national-dna-data-bank-bndg-23 and http://www.mincyt.gob.ar/adjuntos/archivos/000/021/0000021615.pdf and http://servicios.infoleg.gob.ar/infolegInternet/verNorma.do?id=160772).

It was also a kickoff for the creation of the first laboratories that specialized in forensic genetics in Argentina, which emerged in the early 1990s.

Finding that there was a great unmet need for paternity testing, our company incorporated an area of forensic genetics to solve cases of civil as well as criminal causes. Again we faced the lack of commercial kits, hence the genetic markers selection used in parenting studies and our own kit development were based on the scientific literature. We validated studies by means of population surveys, and then communicated in different forensic genetics congresses. There was not any software for statistical calculus, hence all the computations were manually done. "With mathematical formulas written on paper and calculator in hand we calculated Paternity Probabilities…"

## AUTOMATATION

At the end of the 1990s, infectious diseases became the largest field of molecular diagnostics, particularly in viral diseases such as HIV, hepatitis C and B, cytomegalovirus, HPV, and tuberculosis. As a result, large global companies such as Roche Diagnostics (http://www.roche.com.ar/) and Abbott (http://www.abbottlab.com.ar/) settled in Argentina and placed automated equipment for molecular infectious diseases in specialized clinical laboratories. Therefore, specific laboratories performed genetic test for infectious diseases (quantitative and qualitative studies for viruses, viral genotyping studies, etc.), and other laboratories, such as ours, decided to abandon infectious diseases and instead dedicated themselves exclusively to inherited diseases.

In 2001, we rethought our molecular genetic diagnostic laboratory project and founded a new company, Genda S.A. (http://www.genda.com.ar/), which specialized in the diagnosis of genetic diseases and paternity testing, extending molecular studies to risk factors in hemostasis and thrombosis, mitochondrial disease[12] analysis, and panels of higher prevalence diseases. Since 2001, with the appearance of automatic sequencers in the market, techniques could be automated and commercial kits could be used for genetic and paternity testing (Butler, 2011). Automatation led to a bigger number of private laboratories offering genetic diagnosis and paternity testing, resulting in increased competition and a significant studies' price reduction. To illustrate, in 1993 the average value of a paternity test ranged from US$1500 to US$2500, while in 2003 the average value was US$400. Today, in 2017, a paternity test costs US$150.

## HUMAN GENETICS AND HEALTH SYSTEM

Argentina's health coverage system is divided into three parts: the public health system that is free of charge and covers 46% of our population through hospitals distributed throughout the country; the trade union health scheme supported by working class contributions, which covers 46.5% of the population; and the prepaid health care plan system, which covers 7.5% of the population (Penchaszadeh, 2013).

The Compulsory Medical Plan, which lists the practices and diagnoses that any system has to cover 100% of the cost, does not include most of the current genetic tests, leaving the decision to the patient's health service provider, and, until a few years ago, the patients generally had to pay for the study. Fortunately, in recent years this trend is being reversed, showing a growing predisposition of the different health systems to cover the costs of complex genetic test on presentation of a medical history and a budget by the patient that are evaluated by a special audit committee to be paid on special authorization.

Part of the work of diagnostic laboratories has been the promotion of genomic study coverage among the medical community and even among patients' relatives. Today this task is favored by the strong dissemination in social networks of the need for genetic studies and their usefulness.

To date, there is still a lack of professionals working in genetic counseling and prevention in our country. The state has not given priority treatment to these issues. That's why genetics is one of the most defunded fields in medicine. Both public and private genetic centers are concentrated in the city of Buenos Aires, the province of Buenos Aires, and a few cities in the interior of the country. This creates serious information access problems for patients' relatives, pre- and postconception genetic counseling, and the possibilities of having the study costs covered by health systems.

A survey carried out in 2009 at the Garrahan Hospital showed that about 2200 pediatric (nonprenatal) diagnostic studies were performed in the public sector

---

[12] Metabolic diseases due to mutations in the mitochondrial DNA.

per year (Penchaszadeh, 2013). It is estimated that the private sector would have done the same amount of studies. These included the most common genetic diseases: cystic fibrosis, spinal muscular atrophy, fragile X syndrome, Duchenne and Becker muscular dystrophy, Prader Willi syndrome[13], achondroplasia[14] and, hypochondroplasia, Steinert myotonic dystrophy, Friedreich's ataxia, etc.) To date, there is no data available about the frequency of studies that are carried out in the adult population, either for genetic or oncological diseases.

## NEXT-GENERATION SEQUENCING

Patricia dreamed of having children. Her mother had died of breast cancer when she was a child, and her maternal aunt has battled against breast cancer for years. Her geneticist advised her on the possibility of studying her aunt to find out if she had a mutation in any of the 19 genes that are now associated with hereditary breast and/or ovarian cancer. If her aunt was a carrier, they would confirm the suspicion that her family was transmitting hereditary breast cancer, and then Patricia would be studied. After performing the studies, a mutation was found in her aunt. It was then found that Patricia had inherited the genetic mutation that placed her within a women group with a high chance of being exposed to breast cancer. She was about to decide not to have children when she felt that perhaps she was dragging her husband into making the wrong decision. They consulted a specialist who proposed to them that they undergo an embryonic selection treatment. Mutation-free embryos would be implanted in her uterus. The couple accepted the proposal. Patricia gave birth to two girls free of this mutation. Today Patricia does not think about the future. She is calm, because she knows that her daughters can grow to be healthy women and mothers who will see their children grow up and—why not?—their grandchildren, too.

The Human Genome Project, which began in 1990, required 11 years of work to achieve a complete human genome sequence, 20 research groups from different countries that worked together, and billions of dollars.

Since the appearance of next-generation sequencing (NGS)[15] in the market, it has been possible to sequence up to 100 human genomes in less than 2 weeks, and at a cost lower than US$1000.

This has not only caused a revolution in genetic testing, but is also changing the form of research in medicine, giving the possibility of knowing the

---

[13] A Neurological inherited disorder.
[14] Defective conversion of cartilage into bone, especially at the epiphyzes of long bones, producing a type of dwarfism.
[15] NGS also known as high-throughput sequencing, is the catch-all term used to describe a number of different modern sequencing technologies.

causal mechanisms of little-known diseases, understanding tumor genesis, expanding the population genetics field and venturing into the variants that generate risks of complex diseases, carrying out pharmacogenomic studies, improving work methodologies in forensic genetics, etc. (Eichler et al., 2010).

Since 2010 biotech companies worldwide have begun to incorporate the massive amounts of gene sequencing into their diagnostic services and to offer the sequencing of gene panels by using the NGS technique (https://www.illumina.com/content/dam/illumina-marketing/documents/products/illumina_sequencing_introduction.pdf). Panel analysis consists of the simultaneous study of several genes related to a pathology. There are gene panels to find mutations in ataxias, epilepsies, mental retardation, cardiovascular diseases, ophthalmologic, deafness, lung diseases, and cancers, among other pathologies. Exome[16] studies began to be applied in order to look for the origin of unknown or rare diseases.

At that moment, because of the time and high costs that a single-gene sequencing demanded, gene panels through NGS became more profitable and informative, and therefore diagnosis quality and physicians' advice have improved.

The possibilities of offering panel studies in Argentina instead of single-gene sequencing became a new challenge to the medical community culture—they became accustomed to request studies no longer for a single gene, but for gene clusters associated with the same clinical symptomatology, and such information would be construed in that context.

In 2014 the competition among molecular diagnostic laboratories in Argentina was fundamentally based on the most common genetic diseases and paternity testing. Because of a supply cost issue, the use of automated equipment pointed to those studies and not to gene sequencing studies for complex or low-frequency diseases. Genda saw the possibility of offering all those studies that were not developed in the country through agreements and referrals to well-known centers from the United States and Europe (Blueprint, Centogene, Emory, Color, and others). Genda became known abroad as a reference center for Argentina and received numerous representation proposals for diagnostic companies.

Since then, Genda has focused on providing ongoing advice to physicians, informing them about new studies and offering support on genetic information interpretation. We currently receive questions on a great diversity of genetic, neurological, malformation, cardiological, prenatal, and ophthalmological studies. Questions come from patients, physicians, and even health care systems that seek to support a patient's genetic study.

---

[16] The portions of a gene or genome that code information for protein synthesis; the exons in the human genome.

## PERSONALIZED MEDICINE

In 2011 personalized medicine appeared, a new specialty that allows, by taking a DNA sample, one to determine both individual characteristics related to food, nutrition, metabolism and physical exercise, as well as the risk of developing multifactorial diseases[17] such as diabetes or cardiovascular disease (Margaret et al., 2010). In Argentina some companies began to emerge that offered such tests. The circuit consisted of taking a sample from the patient, sending it to the United States for testing, and then receiving the results, which were relayed to the patient by specialized physicians.

Associations between polymorphisms and disease risk have presented scientific evidence through tests implemented by consortia that perform genome wide association studies (http://www.gwascentral.org/). Although its usefulness in disease prevention is still questioned, studies related to nutrition and exercise aimed at improving life quality began to show evidence of improvements in patients who follow the recommendations suggested from the results. Unfortunately, there are still few published studies showing significant differences in the application of these tests in comparison with conventional treatments (prevention guidelines and general population health routine controls), and these are mainly focused on nutrigenetics[18] and pharmacogenetics[19] (Fenech et al., 2011; Scott, 2011).

In Argentina, attempts to introduce these tests so far have not yielded good results, as we try to educate the medical community to understand the meaning of a particular genetic association.

## NONINVASIVE PRENATAL TEST, A NEW STEP IN PRENATAL DIAGNOSIS

In our country at present, the routine monitoring and control of fetal health during pregnancy covered by the three health systems includes fetal nuchal translucency and crown-rump length, values of BHCG or human chorionic gonadotropin in blood, which produces the placenta, the values of another hormone called PAAP-A, prenatal puncture, and chromosome analysis from chorionic villi or amniotic fluid (American College of Obstetricians and Gynecologists, 2007). The latter of these studies is greater for the population over 35 years old, which carries a risk of between 0.5% and 0.2% of pregnancy loss due to the invasive degree of puncture.

---

[17] Diseases due to the interaction between many genes and the environment.
[18] The branch of science concerned with the effect of heredity on diet and nutrition.
[19] The branch of pharmacology that examines the relation of genetic factors to variations in response to drugs.

In 2012, a new way of approaching prenatal studies arose in our country, as two North American companies, Verinata (http://www.illumina.com/) and Natera (http://www.natera.com/), entered the market looking for partnerships with genetics centers to offer noninvasive prenatal studies (NIPT) (Palomaki et al., 2012). These studies using NGS technology are based on the analysis of fetal DNA circulating in the maternal plasma and the search for aneuploidies most frequently associated with advanced maternal age (trisomy 21, 13, 18, X and Y chromosome aneuploidies) with a sensitivity around 99%.

Because of its good reputation in the genetic and diagnosis field, Genda was one of the first centers contacted by these companies to enter the market. The work was arduous, since it was again necessary to convince the obstetric community of the convenience of these studies for their patients, because although the pregnancy period in which these anomalies are detected decrease for 12–10 weeks and do not present risk of pregnancy loss, these studies are not currently covered by the health system and are too costly, only being affordable to people who have high purchasing power. By 2016, about 350 tests/month were carried out in Argentina, distributed among five centers that send the samples to companies abroad. It is surprising how these tests, which tend to guarantee the arrival of a healthy child, had and have an unusual penetrance in the diagnosis market. In addition, as maternity has been postponed throughout Western society, women over 35 choose these tests to know the state of their child's health.

## CURRENT STATUS OF GENOMICS IN ARGENTINA

Although there is a growing demand for genetic studies in the country, and the science and technology policies of the last 10 years have favored the construction of genomic research centers with the latest generation equipment, promoting the creation of the first genomic platform and the first bioinformatic platform of the country, these centers are still at the research stage and have not been massively turned to the application of such platforms in human genetic diagnosis. There are still no genomic databases of our population that can be consulted when analyzing the incidence and impact of a polymorphism or a mutation.

An exploratory survey carried out in 2016 by the National Genomic Data System (www.datosgenomicos.mincyt.gob.ar/pdfs/Encuesta_Exploratoria_SNDG_2016_Resultados.pdf) by 108 research groups from 72 Argentine institutions working on projects involved in the analysis, production, or use of genomic data shows the following:

- In Argentina, mainly in Buenos Aires, Córdoba, Mendoza, and Santa Fe, there are several groups of scientists that use gene or genome data in their research projects.

- So far, the data mostly come from individual genes, although whole genome sequencing is growing.
- Research analyzing genomic data on human diseases accounts for only 5% of total projects.

There are few private centers in Argentina that have NGS technology for genomic studies or that plan to incorporate them. The reasons are basically economic, since the lack of large test requests generates a very high cost per capita (average reagent cost in one panel: US$500). That's why these tools are limited to the public sector and companies have opted to enter into agreements with companies abroad that provide these services at a lower cost.

At present, in Europe and the United States, genetic diagnostic companies are receiving significant investments and growing exponentially. Undoubtedly, since the incorporation of new generation sequencing, the possibility of diagnosing diseases is changing. Today physicians try to find the origin of all diseases, such as cardiovascular disease, malformations, mental retardation, and many others. Argentina cannot compete with these large companies because of, on the one hand, the high cost of machine and reagent and, on the other hand, the low number of each study that is requested. Given this situation, laboratories like ours have signed representations with important international companies. Working with them we are able to solve many patients' diagnoses. From the private sector, we have focused on spreading more awareness of the advances in genetics and diagnosis. We do this through the networks, talks, and specialists who visit physicians to inform them about current advances. At present, we are receiving a large number of budget requests for testing cancer, epilepsies, mental retardation, Marfan syndrome[20], long QT syndrome[21], and many other diseases. To our surprise, the trade union health scheme and the prepaid health care plan have begun to cover a large number of these studies. Although sometimes the process takes a while, the results are proving to be very satisfactory. We think that given the continuous increase in demand, the incorporation of NGS technologies to carry out studies in Argentina is getting closer.

## References

American College of Obstetricians and Gynecologists, 2007. Screening for fetal chromosomal abnormalities. ACOG Practice Bulletin No. 77. Obstet. Gynecol. 109 (1), 217–227.

Baranzini, S., Giliberto, F., Mariana Herrera, García Erro, M., Grippo, J., Szijan, I., 1998. Deletion patterns in Argentine patients with Duchenne and Becker muscular dystrophy. Neurol. Res., 409–413, Col. 20 ISSN: 0161-6412/98.

---

[20] A genetic disorder that affects the body's connective tissue.
[21] A rare congenital and inherited or acquired heart condition in which delayed repolarization of the heart following a heartbeat increases the risk of fainting and sudden death due to ventricular fibrillation.

Butler, J.M., 2011. Advanced Topics in Forensic DNA Typing: Methodology. Available from: https://books.google.com.ar/books?isbn=0123745136.

Chertkoff, L., Visich, A., Bienvenu, T., Grenoville, M., Segal, E., Carniglia, L., Kaplan, J.C., Barreiro, C., 1997. Spectrum of CFTR mutations in Argentine cystic fibrosis patients. Clin. Genet. 51, 43–47.

Eichler, E.E., Flint, J., Gibson, G., Kong, A., Leal, S.M., Moore, J.H., Nadeau, J.H., 2010. Missing heritability and strategies for finding the underlying causes of complex disease. Nat. Rev. Genet. 11, 446–450.

Fenech, M., El-Sohemy, A., Cahill, L., et al., 2011. Nutrigenetics and nutrigenomics: viewpoints on the current status and applications in nutrition research and practice. J. Nutrigenet. Nutrigenom. 4 (2), 69–89.

Iglesias, D.M., Martín, R.S., Fraga, A., Virginillo, M., Kornblihtt, A.R., Arrizurieta, E., Viribay, M., San Millan, J.L., Herrera, M., Bernath, V., 1997. Genetic heterogeneity of autosomal dominant polycystic kidney disease in Argentina. J. Med. Genet. 34 (10), 827–830.

Margaret, A., Hamburg, M.D., Francis, S., Collins, M.D., 2010. The path to personalized medicine. N. Engl. J. Med. 363, 301–304.

Morero, S., Eidelman A., Lichtman, G, 2002. La noche de los bastones largos, 2nd edn. Buenos Aires: Nuevohacer Grupo Editor Latinoamericana. Collection: Colección Temas. ISBN 950-694-684-1.

Palomaki, G.E., Deciu, C., Kloza, E.M., Lambert-Messerlian, G.M., Haddow, J.E., Neveux, L.M., Ehrich, M., Van Den Boom, D., Bombard, A.T., Grody, W.W., Nelson, S.F., Canick, J.A., 2012. DNA sequencing of maternal plasma reliably identifies trisomy 18 and trisomy 13 as well as Down syndrome: an international collaborative study. Genet. Med. 14 (3), 296–305, PMC 3938175. PMID 22281937.

Penchaszadeh, V.B., 2013. Genetic testing and services in Argentina. J. Comm. Genet. 4 (3), 343–354.

Saiki, R., Scharf, S., Faloona, F., Mullis, K., Horn, G., Erlich, H., Arnheim, N., 1985. Enzymatic amplification of beta-globin genomic sequences and restriction site analysis for diagnosis of sickle cell anemia. Science 230 (4732), 1350–1354, PMID 2999980.

Scott, S.A., 2011. Personalizing medicine with clinical pharmacogenetics. Genet. Med. 13 (12), 987–995.

Smith, G., 2005. The genomics age: how DNA technology is transforming the way we live and who we are. AMACOM—Am. Manage. Assoc.

Southern, E.M., 1975. Detection of specific sequences among DNA fragments separated by gel electrophoresis. J. Mol. Biol. 98 (3), 503–517, ISSN 0022-2836. PMID 1195397.

Yang, S., Rothman, R.E., 2004. PCR-based diagnostics for infectious diseases: uses, limitations, and future applications in acute-care settings. Lancet Infect. Dis. 4 (6), 337–348.

## Further Reading

Igarreta, P., Bertolini, R., Herrera, M., Alvarez, F., Bernath, V., 2003. 13-XI Congreso. Panamericano de Neurologia y 58 Congreso Chileno De Neurologia. Santiago de Chile 8-11 de octubre de. SCA1, SCA2, SCA3 AND SCA6 Frequency in argentinean autosomal dominant ataxic patients.

# CHAPTER 8

# Economic Evaluation and Cost-Effectiveness Analysis of Genomic Medicine Interventions in Developing and Emerging Countries

**Anastasios Mpitsakos\*, Christina Mitropoulou\*\*,
Theodora Katsila\*, George P. Patrinos\*,†**

*\*University of Patras School of Health Sciences, Patras, Greece; \*\*The Golden Helix Foundation, London, United Kingdom; †United Arab Emirates University, Al-Ain, United Arab Emirates*

## INTRODUCTION

Pharmacogenomics focuses on the implementation of individual genomic information in the context of clinical care (e.g., for diagnostic or therapeutic decision-making) as well as the health outcomes and policy implications of that clinical use. Today, several clinical studies and data report the pharmacogenomics benefit, especially when patient stratification is needed in various clinical fields, such as those of oncology, cardiology, rare and undiagnosed diseases, and infectious diseases. For this, both the United States Food and Drug Administration (FDA; http://www.fda.gov/) and the European Medicines Agency (EMA; http://www.ema.europa.eu/) have added pharmacogenomics labeling on approved drugs.

One may say that the goal of any health system is to provide high-quality health services to their defined population on an equal basis, to allow quick access to innovation that improves value, to produce a large number of health services to meet the needs of the population, and to do all of this efficiently by consuming as few resources as possible. In most developed and emerging countries, these health system activities are centrally organized through complex systems of political oversight, planning, and financing. Nevertheless, achieving these goals is impeded by certain factors, such as demographic problems, new expensive health technologies, unhealthy lifestyle, medical errors, supply-induced demand for services, and public expectations.

Genomic Medicine in Emerging Economies. http://dx.doi.org/10.1016/B978-0-12-811531-2.00008-4

In view of these factors, in most countries, governments or health insurers have taken initiatives to influence the price and utilization of health care technologies with the use of health technology assessment (economic evaluation) in health care (Drummond et al., 2015). One stated objective of these schemes is to encourage efficiency or cost-effectiveness. In principle, economic evaluation should be relevant to decisions about the pricing and reimbursement of health technologies, as it offers a way of estimating the additional value to society of a new intervention (such as a genetic technology) relative to current therapy. Often the application of economic evaluation in pricing and reimbursement is subjected to changes among countries, but in most cases represent, ideally, a tool for the rational use of limited recourses (Fragoulakis et al., 2015).

Herein, we review economic evaluation and cost-effectiveness data (or their lack of) concerning developing and emerging countries worldwide. Europe will serve as a model to point out that despite recent advances in genomic medicine and health economics, pharmacogenomics is still in its infancy for the majority of low-resourced and developing European countries.

## DEVELOPING AND EMERGING COUNTRIES—A WORLDWIDE PERSPECTIVE

Pharmacogenomics appears to be a useful tool for the formulation of public health decisions in the developing and emerging world, as it attempts to have a beneficial impact on patients via optimum decision-making and disease management. No doubt, to reach its anticipated potential, pharmacogenomics has to overcome current scientific, legal, ethical, political, and economic challenges. For this to occur, an innovative collaboration between various stakeholders, such as health care providers, universities, and nongovernmental and international organizations is required.

Even though emerging countries have been considered appropriate for private investments, much still has to be done in these countries from a social standpoint.

Indeed, an emphasis should be given to improving healthcare delivery infrastructure, to facilitating access to preventive and curative care, and to providing health insurance coverage for a broader population. Indicatively, in India, most of its 1.2 billion population is lacking proper access to healthcare, as only the very wealthy can afford to visit private hospitals, equipped with the latest imaging and medical devices (Rebecchi et al., 2016). Currently, the achievement of universal health coverage in emerging economies is of high priority to the global community to ensure that such countries will sustainably grow on a long-term basis. Some emerging countries, such as China, have made promising progress in that direction, although many are struggling to keep up with a rising population burden (Zhang et al., 2017).

The rational use of healthcare technologies is based on a review of efficacy, safety, and quality of similar agents as well as their comparative efficacy, safety, and their "value-for-money." Frequently, an important deficiency in the process of a national formulary is the lack of national data, which is a common issue in developing and emerging countries. Drug efficacy and/or toxicity data are mostly generated by European/Anglo-American populations, despite the fact that local/national data concerning efficacy, safety, and quality is "mandatory" with regard to healthcare treatment of worldwide communities. Standardization remains an issue, especially when developing and emerging countries are considered, as they are struggling for resources. Thus pharmacogenomics is slowly becoming a part of health care systems in Western high-income countries, as low- and middle-income countries are lacking coherent national policies.

## DEVELOPING AND EMERGING COUNTRIES—A EUROPEAN PERSPECTIVE

Despite the advances in pharmacogenomic research and its use in the clinic in high-income European countries (especially, the United Kingdom, the Netherlands, and Germany), the clinical application of pharmacogenomics is still in its infancy for the majority of low-resourced and developing countries. Notably, the anticipated cost-efficacy outcomes following the implementation of pharmacogenomics in developing and emerging counties are greatly needed, as the public health system or other public sectors (such as education and environmental policies) may be empowered by this surplus.

Many note that there is some criticism inherent in the use of the term "developing country," as the term implies the inferiority of a developing country or an undeveloped country compared with a developed country, and that exact inferiority relates to their economy. Hence, the term "low-income" will replace those of "developing" and "emerging" from now on. According to 2017 data, low-income economies are defined as those with a gross national income (GNI) per capita of US$1025 or less; lower middle-income economies are those with a GNI per capita between US$1026 and US$4035; upper middle-income economies are those with a GNI per capita between US$4036 and US$12,475; high-income economies are those with a GNI per capita of US$12,676 or more. Table 8.1 summarizes the low-income European countries and their corresponding GNI values.

Similar to worldwide challenges when low-income countries are considered, if pharmacogenomics is to be implemented in the clinic and involved in

| Table 8.1 Low-Income European Countries and Their Corresponding GNI Values | |
|---|---|
| **Country** | **GNI** |
| Moldova | 2,220 |
| Ukraine | 2,620 |
| Armenia | 3,880 |
| Kosovo | 3,950 |
| Georgia | 4,160 |
| Albania | 4,290 |
| Bosnia and Herzegovina | 4,680 |
| Former Yugoslav Republic of Macedonia (FYROM) | 5,140 |
| Serbia | 5,500 |
| Belarus | 6,460 |
| Azerbaijan | 6,560 |
| Bulgaria | 7,220 |
| Montenegro | 7,240 |
| Turkey | 9,950 |
| Russia | 11,400 |
| Kazakhstan | 11,580 |
| Romania | 12,670 |

decision-making in public healthcare policy, its cost-effectiveness must be determined. In short, cost-effectiveness analysis (CEA) is a form of economic analysis that compares the relative costs and outcomes (effects) of different courses of action (Fragoulakis et al., 2015). Cost-effectiveness analysis is distinct from cost-benefit analysis, which assigns a monetary value to the measure of effect. In the context of pharmacoeconomics, the cost-effectiveness of a therapeutic or preventive intervention is the ratio of the cost of the intervention to a relevant measure of its effect. Cost refers to the resource expended for the intervention, usually measured in monetary terms, such as US dollar, British pound, and so on. The measure of effects depends on the intervention being considered. A special case of CEA is the cost-utility analysis, where the effects are measured in terms of years of full health lived, using a measure such as quality-adjusted life years (QALY) or disability-adjusted life years. Cost-effectiveness is typically expressed as an incremental cost-effectiveness ratio (ICER), the ratio of change in costs to the change in effects.

Today, cost-effectiveness studies have mostly been done from the perspective of high-income countries, even though an argument can be made that the analyses could be much more important to low-income countries where the consequence of expending scarce resources on technology that is not cost-effective has a much higher opportunity cost. Unfortunately, there is also a shortage of health economists and modelers in low-income countries. Apart from the differences in current drug prices and resource utilization in different countries, another important parameter to determine the cost-effectiveness of a certain medical intervention in different healthcare systems is the variable frequencies of the pharmacogenomic biomarkers. As such, one should bear in mind that a pharmacogenomics-guided medical intervention that is not cost-effective in a certain country may be cost-effective in another country, even if no significant cost differences exist between these two countries, but essentially because of the higher frequency of a pharmacogenomic biomarker in the general population. Notably, in several cases the genomic composition of a population may be different from one geographical region to another (Mizzi et al., 2016). Indicatively, the Russian population clusters into several large ethnogeographical groups, namely Slavs, Northern Caucasus populations, the Finno-Ugric people of north European and Volga-Ural regions, the populations of South Siberia and Central Asia, and the populations of Eastern Siberia and North Asia. For example, the *CYP2D6\*7* allele frequency is significantly higher in the Maltese population compared with the Caucasian average, and the same is true for the *CYP2C9\*3* allele frequency in the Serbian population (Mizzi et al., 2016), suggesting that there may be significant implications in the cost-effectiveness of, for example, risperidone and warfarin treatments, respectively, in these two countries. Another example is screening for the *HLA-B\*1502* allele, which elevates patients' risk of developing Stevens-Johnson syndrome (SJS) and toxic epidermal necrolysis (TEN), when treated with the antiepileptic drugs carbamazepine and phenytoin. A recent systematic review and meta-analysis including 16 studies found considerable variation among different racial/ethnic populations in the relationship between *HLA-B\*1502* and carbamazepine-induced SJS and TEN as illustrated by a summary odds ratio of 79.84 (95% CI, 28.45–224.06) with the following strong relationships warranting screening for three racial/ethnic subgroups: Han Chinese 115.32 (18.17–732.13), Thais 54.43 (16.28–181.96), and Malaysians 221.00 (3.85–12,694.65) (Tangamornsuksan et al., 2013). Among individuals of white or Japanese race/ethnicity, no patients with SJS or TEN were carriers of the *HLA-B\*1502* allele. An intra-country example is a CEA of three ethnic groups in Singapore with different allele frequencies of *HLA-B\*1502*. Genotyping for *HLA-B\*1502* and providing alternate antiepileptic drugs to those who test positive was found to be cost-effective for Singaporean Chinese and Malays, but not for Singaporean Indians based on ICER ratios of US$37,030/QALY for Chinese patients, US$7,930/QALY for Malays, and Us$136,630/QALY for Indians.

# EXAMPLES OF ECONOMIC EVALUATION AND COST-EFFECTIVENESS ANALYSIS

## Anticoagulation Therapy

Warfarin is one of the most studied drugs in Europe, and one of the few drugs in which the cost-effective data has been shared worldwide; because of this it may serve as a comparison tool. In a recent economic evaluation study involving elderly atrial fibrillation patients undergoing warfarin treatment, it was shown that not only did 97% of elderly Croatian patients with atrial fibrillation belonging to the pharmacogenomics-guided group not have any major complications, compared with 89% in the control group, but, most importantly, the ICER of the pharmacogenomics-guided versus the control groups was calculated to be just €31,225/QALY (Mitropoulou et al., 2015). These data suggest pharmacogenomics-guided warfarin treatment represents a cost-effective therapy option for the management of elderly patients with atrial fibrillation in Croatia, which may very well be the case for the same and other anticoagulation treatment modalities in neighboring countries.

A review on the global differences among healthcare systems and costs regarding *CYP2C9* and *VKORC1* genotyping-guided coumarin derivatives treatment took a closer look at anticoagulant care management and its cost in the United Kingdom, Sweden, the Netherlands, Greece, Germany, and Austria. As it was reported, variations between countries were found in the setting of the international normalized ratio (INR) monitoring and coumarin dosing, the frequency of INR monitoring, and in the prevalence of coumarin use. Differences were also found in the quality of anticoagulation, in terms of the percentage of time spent in the target INR range and the rate of complications. Efficacy and cost-effectiveness of genotyping prior to treatment can be influenced by the management and quality of anticoagulant care. In countries where anticoagulant care is less well organized, there is the highest probability for pharmacogenomics to be cost-effective. Nevertheless, genotyping might still be a cost-effective strategy in countries where anticoagulant care is well organized, as less INR measurements would be required when patients reach a stable dose early on with genotyping. Genotyping costs and effects should be taken into full consideration on the basis of data impact, as reported by Meckley et al. (2010), whose policy model suggested a small clinical benefit for warfarin pharmacogenomics testing, yet with significant uncertainty in economic value (Meckley et al., 2010). Because of significant uncertainties regarding important assumptions in their Markov decision analytic model, Verhoef et al. (2013) stated that it was too early to conclude whether or not Dutch patients with atrial fibrillation starting phenprocoumon should be genotyped, even though pharmacogenetic-guided dosing of phenprocoumon had the potential to increase health slightly in a cost-effective way. The main factors for this uncertainty were the effectiveness of a pharmacogenetic-guided dosing regimen as well as the costs of the genetic test. A couple of years later, Verhoef et al. (2015) investigated the cost-effectiveness of a pharmacogenetic dosing algorithm versus a clinical dosing algorithm for phenprocoumon and acenocoumarol versus clinical dosing in the Netherlands. Pharmacogenetic dosing was reported to increase costs by €33 and QALYs by 0.001, and, as such, improve health only slightly when compared with clinical dosing. For phenprocoumon, the ICER was €28,349 per QALY gained; for acenocoumarol, €24,427 per QALY gained was concluded. Even though, at a willingness-to-pay threshold of €20,000 per QALY, the pharmacogenetic dosing algorithm was not likely to be cost-effective compared with the clinical dosing algorithm, the authors stated that availability of low-cost genotyping would make it a cost-effective option. For typical patients with nonvalvural atrial fibrillation, warfarin-related genotyping was unlikely to be cost-effective, yet might be cost-effective in those patients being at high hemorrhagic risk. Of particular interest is the study of Jowett et al. (2008) that focused on the time and traveling costs that patients incur to themselves and society in order to attend anticoagulation clinics, especially when taking into account that therapy success requires regular monitoring and, frequently, dose adjustment. Patients were found to incur considerable costs when visiting anticoagulation clinics, and these costs varied by country, ranging from €6.9 (France) to €20.5 (Portugal) per visit. No doubt, a broad economic perspective becomes of fundamental importance when considering the cost-effectiveness of warfarin (Verhoef et al., 2012).

A few studies have investigated alternatives to warfarin for stroke prophylaxis in patients with atrial fibrillation, raising the interesting question of whether these alternatives are cost-effective (Pink et al., 2014). On the basis of the results from randomized evaluation of long-term anticoagulation therapy (RE-LY) (Wallentin et al., 2010) and other trials, a decision-analysis model was developed to compare the cost and quality-adjusted survival of various antithrombotic therapies (dabigatran, aspirin, and warfarin) (Shah and Gage, 2011). A Markov model was run in a hypothetical cohort of 70-year-old patients with atrial fibrillation, using a cost-effectiveness threshold of $50,000/QALY. Dabigatran 150 mg (twice daily) was found to be cost-effective in patient populations at high risk of hemorrhage or stroke, unless INR control with warfarin was excellent. Warfarin was cost-effective in moderate-risk patient populations, unless INR control was poor. Nevertheless, neither dabigatran nor rivaroxaban were cost-effective options when relative risks of clinical events served as inputs to an economic analysis, following a clinical trial simulation of warfarin. Along the cost-effectiveness frontier, apixaban was the most cost-effective treatment. Interestingly enough, O'Brien and Gage (2005) compared quality-adjusted survival and cost among ximelagatran, warfarin, and aspirin for patients with chronic atrial fibrillation. According to their Semi-Markov decision model findings and assuming equal effectiveness in stroke prevention and decreased hemorrhage risk, ximelagatran was not likely to be cost-effective in patients with atrial fibrillation, unless they had a high risk of intracranial hemorrhage or a low quality of life with warfarin. On the basis of head-to-head evidence from randomized controlled trials, the cost-utility of eprosartan versus enalapril (primary prevention) and versus nitrendipine (secondary prevention) has also been investigated. For this, the HEALTH model (Health Economic Assessment of Life with Teveten for Hypertension) was used (Schwander et al., 2009). When a €30,000 willingness-to-pay threshold per QALY gained was considered, eprosartan was cost-effective in both primary (patients ≥50 years old and a systolic blood pressure ≥160 mm Hg) and secondary prevention (all investigated patients).

## Antidepressants

Despite the case of coumarin derivates, the economic evaluation regarding the use of pharmacogenetic tests to guide treatment in major depression cases in high- and middle-income European countries showed that pharmacogenetic tests may be cost-effective in the high-income countries of Western Europe, but not in the middle-income countries of Eastern Europe (Olgiati et al., 2012). The study described the cost-utility of incorporating 5-HTTLPR genotyping prior to drug treatment in the case of a major depressive disorder. The drugs citalopram or bupropion were selected, based on the response and tolerability predicted by the 5-HTTLPR profile or standard treatment guidelines. The model was constructed for the European regions with high gross domestic product (GDP) (Austria, Belgium, Cyprus, the Czech Republic, Denmark, Finland, France, Germany, Greece, Ireland, Italy, Luxembourg, Malta, the Netherlands, Portugal, Slovenia, Sweden, and the United Kingdom), middle GDP (Bulgaria, Poland, Romania, and Slovakia) and middle-high GDP (Estonia, Hungary, Latvia, and Lithuania). The results showed that in Eastern Europe a larger proportion of individuals were treated in inpatient facilities, and hence the overall treatment cost was higher than that in Western Europe, although the costs for the single service were lower. However, the results also showed the same incremental cost for the pharmacogenetic approach in all the European regions tested (ICER for regions with high GDP was $1147; ICER for regions with middle GDP was $1158; ICER for regions with middle-high GDP was $1179). The authors suggested that the critical factor that determined whether the pharmacogenetic test was cost-effective depended on the threshold, which was proportional to the economical level of the county. The World Health Organization indicates an intervention as highly cost-effective if ICER is inferior or equal to the GDP per capita, and cost-effective if the ICER is between one and three times the GDP per capita (http://www.who.int/choice/costs/CER_thresholds/en). As data on costs were from 2009, the current economic situation in the countries considered might be significantly different now. The presented cost-utility model was robust against the variations in all the parameters except for the cost of the genetic test, which produced the greatest changes in ICER. It was suggested that as long as genetic analysis is an expensive procedure, its applicability is limited to the richest areas in the Eurozone.

## Allopurinol

In the Thai population, the cost-utility analysis of *HLA-B\*5801* testing before the allopurinol administration was performed using a societal perspective (Saokaew et al., 2014). The model compared the cost-utility of a hypothetical pharmacogenetics treatment arm to conventional treatment. The carriers of the *HLA-B\*5801* allele received the alternative drug (probenecid), whereas all the others received allopurinol. All probability data were derived from the literature (Somkrua et al., 2011) or calculated from conditional probabilities. Cost-effectiveness was sensitive to the medical care cost of gout management, the incidence of allopurinol-induced SJS/TEN, the probability of death with SJS/TEN, and the cost of genetic testing. The authors suggested that the implementation of genotypic testing for the *HLA-B\*5801* allele might be potentially cost-effective only in countries with a large number of subjects at risk, such as those from the Southeast Asian and Han Chinese countries or those with a high prevalence of the *HLA-B\*5801* allele. *HLA-B\*5801* allele prevalence was reported as high as 5.5–15% in the Thai and Han Chinese populations, whereas it was as low as 0.6% and 0.8% in the Japanese and European populations. The findings of a Korean study performed on a hypothetical cohort of gout patients with chronic renal insufficiency were similar (Park et al., 2015). This study took a national health payer's perspective and concluded that allopurinol treatment based on *HLA-B\*5801* genotyping could be more cost-effective, resulting in a better outcome than the conventional treatment strategy.

## Human Papillomavirus Testing

In South Africa a model was developed to determine the cost-effectiveness of several cervical cancer screening strategies via human papillomavirus (HPV) testing (Vijayaraghavan et al., 2009). Screening strategies included conventional cytology, cytology followed by HPV testing for triage of equivocal cytology, HPV testing, HPV testing followed by cytology for triage of HPV-positive women, and coscreening with cytology. Primary outcome measures included QALY saved, ICERs, and lifetime risk of cervical cancer. Results showed that screening once every 10 years reduced the lifetime risk of cervical cancer by 13%–52%, depending on the screening strategy used. When strategies were compared incrementally, cytology with HPV triage was less expensive and more effective than screening using cytology alone. HPV testing with the use of cytology triage was a more effective strategy, with a cost of an additional R42,121 (R represents the local currency) per QALY. HPV testing with colposcopy for HPV-positive women was the next most effective option. Simultaneous HPV testing and cytology coscreening was the most effective strategy and had an incremental cost of 25,414 rand (local currency) per QALY. Conclusively, HPV testing for cervical cancer screening was a cost-effective strategy in South Africa.

In an economic evaluation which was conducted in Lebanon, researchers evaluated cytology and HPV DNA screening for women aged 25–65 years, with varying coverage from 20% to 70% and frequency from 1 to 5 years (Sharma et al., 2017). In Lebanon, there is no national organized cervical cancer (CC) screening program, and thus screening is limited only to those women who can pay out of pocket. Hence, the study evaluated the cost-effectiveness of increasing screening coverage and extending intervals. The model was calibrated to epidemiological data from Lebanon, including CC incidence and HPV type distribution. Results showed that at 20% coverage, annual cytologic screening reduced lifetime CC risk by 14% and had an ICER of I$80,670/year of life saved, far exceeding Lebanon's GDP per capita (I$17,460), a commonly cited cost-effectiveness threshold. By comparison, increasing cytologic screening coverage to 50% and extending screening intervals to 3 and 5 years provided a greater CC reduction (26.1% and 21.4%, respectively) at lower costs, when compared with 20% coverage with annual screening. Screening every 5 years with HPV DNA testing at 50% coverage provided greater CC reductions than cytology at the same frequency (23.4%) and was cost-effective, assuming a cost of I$18 per HPV test administered (I$12,210/year of life saved); HPV DNA testing every 4 years at 50% coverage was also cost-effective at the same cost per test (I$16,340). Increasing coverage of annual cytology was not found to be cost-effective. The analysis concluded that the current practice of repeated cytology in a small percentage of women is inefficient, whereas increased coverage to 50% with extended screening intervals provides greater health benefits at a reasonable cost and thus can more equitably distribute health gains. Notably, novel HPV DNA strategies offer greater CC reductions and may be more cost-effective than cytology.

In another economic evaluation, the cost-effectiveness of different cervical screening strategies was estimated in the Islamic Republic of Iran, a Muslim country with a low incidence rate of invasive CC (Nahvijou et al., 2016). An 11-state Markov model was constructed, in which the parameters included regression and progression probabilities, test characteristics, costs, and utilities; these were extracted from the primary data and current literature. Comparing strategies included Pap smear screening, HPV DNA testing, plus Pap smear triaging with various starting ages and screening intervals. Model outcomes included lifetime costs, life years gained, QALY and ICERs. It was concluded that the prevented mortalities for the 11 strategies compared with no screening varied from 26% to 64%. The most cost-effective strategy was HPV screening, starting at the age of 35 years and repeated every 10 years. The ICER of this strategy was US$8875 per QALY compared with no screening. Screening at 5-year intervals was found to be cost-effective on the basis of GDP per capita in Iran. Overall, cervical screening with HPV DNA testing for women in Iran, beginning at the age of 35 and repeated every 5 or 10 years was a cost-effective and recommended treatment.

# CONCLUSION AND FUTURE PERSPECTIVES

The above examples suggest that pharmacogenomic economic evaluation studies must be replicated in every country to inform policymakers prior to the implementation of a pharmacogenomic-guided medical intervention, following the evaluation of its cost-effectiveness based on characteristics specific to each country. This is a daunting expectation given the rapidly increasing number of pharmacogenomics guidelines of developed countries' regulatory agencies, such as the US FDA, and multinational organizations, such as the Clinical Pharmacogenetics International Consortium, not to mention that many developing countries do not have the necessary resources and expertise to perform the analyses. This issue could be minimized by the construction of generic economic models that would allow input of certain key variables, such as allele frequency and test and treatment costs that vary by country. This model could be used by less experienced individuals who have access to the country-specific variables to generate a first approximation of cost-effectiveness, allowing prioritization between different emerging tests. The decrease of genotyping costs for the once-in-a-lifetime determination of an individual's personalized pharmacogenomics profile, using next generation sequencing technologies including whole genome sequencing, would gradually result in a cost-effective pharmacogenomics-guided treatment for drugs bearing pharmacogenomic testing recommendations on their labels, as the single test could provide information that would inform prescribing choices over the lifetime of a patient.

In addition, the application of population pharmacogenomics would contribute toward making pharmacogenomics a useful tool, which could be incorporated into the clinical practice of every country. In particular, to close the disparity in practical use of pharmacogenomics in the developing world, such approaches (e.g., the Euro-PGx project) would help toward (1) enhancing the understanding of pharmacogenomics in the developing world, (2) providing guidelines for medication prioritization, and (3) building infrastructure for future pharmacogenomic research studies. Such projects would serve to enhance the understanding of pharmacogenomics globally. Although the Euro-PGx project, coordinated by the Golden Helix Foundation (see also Chapter 9), only represents the European continent, officially launched in 2010 and successfully completed in 2016 (Mizzi et al., 2016), it provides the basis for replication in individual countries; similar projects are already about to be launched in other geographical regions, such as the 1000 Pharmacogenes project in Southeast Asia, under the umbrella of the Golden Helix Foundation and the Global Genomic Medicine Collaborative. Other initiatives, such as the Human Heredity and Health in Africa (H3Africa), the Qatar Genome Project, and the Mexico National Institute of Genomic Medicine (INMEGEN) aimed to address such issues through capacity building and empowerment of local researchers to spark a paradigm shift.

In essence, conducting basic research to discover important advances in the applications of genetics, as well as economic analysis to determine the financial consequences of these adoptions, is important in order to ensure that patients receive an acceptable level of care while effectively managing health care resources. As the benefits of genetics have been partially acknowledged by general practitioners, but also remain under consideration, similar scientific projects, such as those mentioned above, must be continued to give further insights into these healthcare technologies.

## Acknowledgments

This work was endorsed by the Genomic Medicine Alliance Health Economics Working Group.

## References

Drummond, M.F., Sculpher, M.J., Claxton, K., Stoddart, G.L., Torrance GW, 2015. Methods for the Economic Evaluation of Health Care Programmes. Oxford University Press.

Fragoulakis, V., Mitropoulou, C., Williams, M.C., Patrinos, G.P., 2015. Economic Evaluation in Genomics Medicine. Elsevier/Academic Press.

Jowett, S., Bryan, S., Mahé, I., Brieger, D., Carlsson, J., Kartman, B., et al., 2008. A multinational investigation of time and traveling costs in attending anticoagulation clinics. Value Health 11, 207–212.

Meckley, L.M., Gudgeon, J.M., Anderson, J.L., Williams, M.S., Veenstra, D.L., 2010. A policy model to evaluate the benefits, risks and costs of warfarin pharmacogenomic testing. Pharmacoeconomics 28, 61–74.

Mitropoulou, C., Fragoulakis, V., Bozina, N., Vozikis, A., Supe, S., Bozina, T., et al., 2015. Economic evaluation of pharmacogenomic-guided warfarin treatment for elderly Croatian atrial fibrillation patients with ischemic stroke. Pharmacogenomics 16, 137–148.

Mizzi, C., Dalabira, E., Kumuthini, J., Dzimiri, N., Balogh, I., Bașak, N., et al., 2016. A European spectrum of pharmacogenomic biomarkers: implications for clinical pharmacogenomics. PLoS One 11, e0162866.

Nahvijou, A., Daroudi, R., Tahmasebi, M., Hashemi, F.A., Hemami, M.R., Sari, A.A., et al., 2016. Cost-effectiveness of different cervical screening strategies in Islamic Republic of Iran: a middle-income country with a low incidence rate of cervical cancer. PLoS One 11, e0156705.

O'Brien, C.L., Gage, B.F., 2005. Costs and effectiveness of ximelagatran for stroke prophylaxis in chronic atrial fibrillation. JAMA 293, 699–706.

Olgiati, P., Bajo, E., Bigelli, M., De Ronchi, D., Serretti, A., 2012. Should pharmacogenetics be incorporated in major depression treatment? Economic evaluation in high-and middle-income European countries. Prog. Neuropsychopharmacol Biol. Psychiatry 36, 147–154.

Park, D.J., Kang, J.H., Lee, J.W., Lee, K.E., Wen, L., Kim, T.J., et al., 2015. Cost-effectiveness analysis of HLA-B5801 genotyping in the treatment of gout patients with chronic renal insufficiency in Korea. Arthritis Care Res. 67, 280–287.

Pink, J., Pirmohamed, M., Lane, S., Hughes, D., 2014. Cost-effectiveness of pharmacogenetics-guided Warfarin therapy vs. alternative anticoagulation in atrial fibrillation. Clin. Pharmacol. Ther. 95, 199–207.

Rebecchi, A., Gola, M., Kulkarni, M., Lettieri, E., Paoletti, I., Capolongo, S., 2016. Healthcare for all in emerging countries: a preliminary investigation of facilities in Kolkata, India. Ann. Ist Super Sanita 52, 88–97.

Saokaew, S., Tassaneeyakul, W., Maenthaisong, R., Chaiyakunapruk, N., 2014. Cost-effectiveness analysis of HLA-B* 5801 testing in preventing allopurinol-induced SJS/TEN in Thai population. PLoS One 9, e94294.

Schwander, B., Gradl, B., Zöllner, Y., Lindgren, P., Diener, H.-C., Lüders, S., et al., 2009. Cost-utility analysis of eprosartan compared to enalapril in primary prevention and nitrendipine in secondary prevention in Europe–the HEALTH model. Value Health 12, 857–871.

Shah, S.V., Gage, B.F., 2011. Cost-effectiveness of dabigatran for stroke prophylaxis in atrial fibrillation. Circulation, 110.985655.

Sharma, M., Seoud, M., Kim, J.J., 2017. Cost-effectiveness of increasing cervical cancer screening coverage in the Middle East: an example from Lebanon. Vaccine 35, 564–569.

Somkrua, R., Eickman, E.E., Saokaew, S., Lohitnavy, M., Chaiyakunapruk, N., 2011. Association of HLA-B* 5801 allele and allopurinol-induced Stevens Johnson syndrome and toxic epidermal necrolysis: a systematic review and meta-analysis. BMC Med. Genet. 12, 118.

Tangamornsuksan, W., Chaiyakunapruk, N., Somkrua, R., Lohitnavy, M., Tassaneeyakul, W., 2013. Relationship between the HLA-B* 1502 allele and carbamazepine-induced Stevens-Johnson syndrome and toxic epidermal necrolysis: a systematic review and meta-analysis. JAMA Dermatol. 149, 1025–1032.

Verhoef, T.I., Redekop, W.K., De Boer, A., Maitland-Van Der Zee, A.H., group, E.-P., 2015. Economic evaluation of a pharmacogenetic dosing algorithm for coumarin anticoagulants in the Netherlands. Pharmacogenomics 16, 101–114.

Verhoef, T.I., Redekop, W.K., Van Schie, R.M., Bayat, S., Daly, A.K., Geitona, M., et al., 2012. Cost-effectiveness of pharmacogenetics in anticoagulation: international differences in healthcare systems and costs. Pharmacogenomics 13, 1405–1417.

Verhoef, T.I., Redekop, W.K., Veenstra, D.L., Thariani, R., Beltman, P.A., Van Schie, R.M., et al., 2013. Cost-effectiveness of pharmacogenetic-guided dosing of phenprocoumon in atrial fibrillation. Pharmacogenomics 14, 869–883.

Vijayaraghavan, A., Efrusy, M., Lindeque, G., Dreyer, G., Santas, C., 2009. Cost effectiveness of high-risk HPV DNA testing for cervical cancer screening in South Africa. Gynecol. Oncol. 112, 377–383.

Wallentin, L., Yusuf, S., Ezekowitz, M.D., Alings, M., Flather, M., Franzosi, M.G., et al., 2010. Efficacy and safety of dabigatran compared with warfarin at different levels of international normalised ratio control for stroke prevention in atrial fibrillation: an analysis of the RE-LY trial. Lancet 376, 975–983.

Zhang, M., Wang, W., Millar, R., Li, G., Yan, F., 2017. Coping and compromise: a qualitative study of how primary health care providers respond to health reform in China. Hum. Resour. Health 15, 50.

# CHAPTER 9

# Raising Genomics Awareness Among the General Public and Educating Healthcare Professionals on Genomic Medicine

**George P. Patrinos**

*University of Patras School of Health Sciences, Patras, Greece; United Arab Emirates University, Al-Ain, United Arab Emirates*

## INTRODUCTION

Genomic medicine aims to make use of an individual's genomic information in the context of guiding the clinical decision-making process. In recent years, there have been significant advances in genomics research that have led to the delineation of the underlying genomic variants, from single nucleotide variations to complex genome rearrangements, and/or identify the altered gene expression patterns with disease pathogenicity. These findings have the potential to guide physicians in their task of estimating disease risk as well as individualizing treatment modalities, which creates unprecedented opportunities for the customization of patient care including the personalization of conventional therapeutic interventions in various medical specialties, mainly oncology, cardiology, and psychiatry (reviewed in Squassina et al., 2010).

However, such developments have yet to make their way into the clinic, one of the most fundamental hurdles being the lack of genomics education by health care professionals, who are insufficiently trained to engage with the delivery of genomics services. At the same time, despite the constant hype about the future potential of personalized medicine and the generally inflated expectations as to the likely health benefits that derive from the genomic revolution, patients and the general public have a very low genetic literacy, which impairs their capacity to meaningfully integrate genomic information into their lives and health care decision-making. This phenomenon is more profound in developing countries and low-resource environments. To better address the latter bottlenecks, activities addressed toward enriching genetics education and raising genomics awareness should be planned and meticulously implemented in a harmonized manner worldwide.

Genomic Medicine in Emerging Economies. http://dx.doi.org/10.1016/B978-0-12-811531-2.00009-6

In this chapter the overall educational and outreaching activities of the Golden Helix Foundation will be outlined, as a paradigm of a well-orchestrated effort to (1) enrich genomics education among healthcare professionals, namely, physicians and biomedical scientists, and (2) increase genomic literacy among patients and the general public.

## GENOMICS EDUCATION OF HEALTHCARE PROFESSIONALS

One of the key elements to ensure efficient integration of genomics into clinical practice is the enrichment of health professionals' genomics knowledge. Presently, a typical practicing physician has only very limited training in genetics, suggested by previous studies in the United States (Haga et al., 2012a; Stanek et al., 2012) and Europe (Mai et al., 2014, and references therein). Also, as far as pharmacogenomic testing is concerned, even when clinical geneticists and genetic counselors were surveyed, only a small fraction of them (e.g., 12% of genetic counselors and 41% of clinical geneticists) indicated that they had ordered or coordinated patient care for pharmacogenomic testing. In the latter case, despite the fact that almost all respondents had received some education on pharmacogenomics, only 28% of counselors and 58% of clinical geneticists indicated that they felt well informed (Haga et al., 2012b). These findings have important implications for the education of physicians and even geneticists, who have a critical role to play in pharmacogenomics testing.

Health professionals' lack of genetics knowledge coupled with the lack of awareness of both patients and the general public about the benefits of genetic information (see below), leaves the public confused as to which test might be beneficial in their own particular case (Patrinos et al., 2013). In a recent survey of Greek pharmacists, it became evident that pharmacists do not feel sufficiently competent to explain the results of pharmacogenomic tests to their clients (Mai et al., 2014), further highlighting the need for pharmacists to receive some basic training about the genetic tests that they sell in their practices.

In order to tackle their self-declared lack of appropriate genomics education, health care professionals must be encouraged (or even obliged by their respective professional bodies) to pursue continuous genetics education, possibly provided in the form of accredited seminars organized by local universities and/or international organizations. Such continuous genetics education would be very important for those healthcare professionals who did not acquire the necessary level of genetics expertise during their undergraduate studies.

At the same time, participation in, often continuous medical education (CME)-accredited, genetics/genomics conferences would also contribute to improving healthcare professionals' genomics education (Kampourakis et al., 2014a,b). To this end, there are a plethora of genomics conferences that are organized in various countries worldwide. However, registration and tuition fees to attend these conferences, which are mostly organized in the United States and Western Europe, are prohibitively expensive for participants originating from developing countries and low-resource environments. As such, there was an urgent need to establish a series of conferences in genomic medicine and pharmacogenomics, where the high scientific quality is met with affordable registration fees and capable of attracting participants from the developing world.

## THE CONCEPT

As previously mentioned, the majority of the prestigious genomics conferences are organized in the United States and Western Europe, which often poses financial and other (e.g., logistics) restrictions to participants from Southeast Asia, the Middle East, and low-resource environments from South and Central Europe, Africa, and Latin America. Ideally, if high-profile academics and researchers were attracted, as plenary and keynote speakers to genomics conferences organized in these regions (which are often characterized by beautiful surroundings and rich history), this would, in turn, attract corporate and charitable entities as sponsors of these scientific events, which would reciprocally reduce the overall registration fees, hence making these meetings more affordable to participants for developing countries. Also, organizing such conferences in different venues and with a different, but always related to genomic medicine, theme each time, would reduce the logistical burden and make it easier and more cost-effective for participants from developing countries to travel to. All the above gave rise to the concept of the Golden Helix conferences that are outlined below.

## THE GOLDEN HELIX CONFERENCES

The Golden Helix conferences are named after the house of Francis Crick ("the Golden Helix"; 19/20 Portugal Place, Cambridge, UK) to emphasize their focus on human genomics and personalized medicine. The themes of these conferences revolve around the fields of genomics, pharmacogenomics, and personalized medicine. There are three different types of Golden Helix conferences:

1. Golden Helix Symposia
2. Golden Helix Pharmacogenomics Days
3. Golden Helix Summer Schools

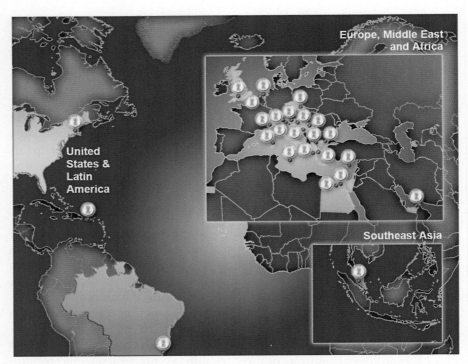

**FIGURE 9.1** Depiction of the various locations worldwide, where Golden Helix conferences have been organized.

The vast majority of Golden Helix conferences have been organized, since 2008, mostly in major cities of developing countries in Europe, the Middle East, Asia, and Africa, attracting very high profile academics and researchers, and large corporate entities (such as pharmaceutical and biotechnology companies), as well as charitable entities as sponsors. The above have significantly reduced the overall registration fees to attend the meetings, organized each time in different venues and with a different, but always related to genomic medicine, theme. In addition, the management of these scientific events, undertaken centrally by the Golden Helix Foundation (www.goldenhelix.org), has reduced the logistical burden and created a recognizable brand of excellence around these series of conferences (Fig. 9.1).

As of November 2008, more than 35 Golden Helix conferences have been organized in 28 different cities and 16 different countries worldwide (Fig. 9.2).

The features of the different Golden Helix conferences are outlined in the paragraphs below.

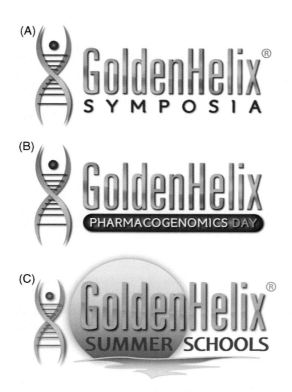

**FIGURE 9.2 Logos of the three different types of Golden Helix conferences.**
(A) Golden Helix Symposia; (B) Golden Pharmacogenomics Days, and (C) Golden Helix Summer Schools.

## Golden Helix Symposia

The Golden Helix Symposia (http://symposia.goldenhelix.org/) are high-caliber international research meetings. The themes of these symposia revolve around the fields of genomic and personalized medicine. In particular, the themes of these symposia series are selected by the Golden Helix Foundation International Scientific Advisory Council, in close partnership with the local organizing committee, so that new and often ground-breaking science is presented at every meeting. Symposia range in length from 2 to 4 days, and average attendance is 450 participants (ranging from 250 to 900 participants), mainly staff scientists and biomedical researchers; postdoctoral fellows and students from academia, working in the field of personalized research; and scientists from pharmaceutical and biotechnology industries, government agencies, and private and research foundations from various countries worldwide. Conference venues are usually prestigious locations in major cities or summer retreats in the southern Mediterranean or eastern regions with

**Table 9.1** Past Golden Helix Symposia

| A/A | Year | Date | City | Country |
|-----|------|------|------|---------|
| 1 | 2008 | November 28–29 | Athens | Greece |
| 2 | 2009 | October 15–17 | Athens | Greece |
| 3 | 2010 | December 1–4 | Athens | Greece |
| 4 | 2012 | April 18–21 | Turin | Italy |
| 5 | 2013 | November 17–19 | Dubai | United Arab Emirates |
| 6 | 2014 | October 30–November 1 | Belgrade | Serbia |
| 7 | 2015 | March 11–13 | Kuala Lumpur | Malaysia |
| 8 | 2016 | January 14–16 | Mansoura | Egypt |
| 9 | 2017 | April 27–29 | Athens | Greece |

beautiful surroundings and rich history. In these symposia, scientific excellence is ensured, as important collaborations are formed, careers are enhanced, and new scientific data are presented.

Previous Golden Helix Symposia have focused on copy number variation and genomic alterations in health and disease (Athens, Greece, 2008 [Le Caignec and Redon, 2009; Patrinos and Petersen, 2009]), pharmacogenomics (Athens, Greece, 2009 [Patrinos and Innocenti, 2010]; Kuala Lumpur 2015), genomic medicine (Athens, Greece, 2010 [Patrinos et al., 2011]; Turin, Italy, 2012 [Kricka and Di Resta, 2013]; Dubai, United Arab Emirates, 2013 [Fortina et al., 2014]), genomics of rare diseases (Belgrade, Serbia, 2014), and of neurodegenerative disorders (Mansoura, Egypt, 2016). Also, the 2017 Golden Helix Symposium has been jointly organized with the third Global Genomic Medicine Collaborative Conference in Athens, Greece, focusing on the implementation of genomic medicine in the clinic (Table 9.1).

In several cases, the Golden Helix Foundation provided fellowships to undergraduate and postgraduate students from low- to medium-income countries to defray registration, and often, accommodation and travel costs.

## Golden Helix Pharmacogenomics Days

The Golden Helix Foundation has also organized, since early 2009, the Golden Helix Pharmacogenomics Days, an international educational event series organized in major cities with large academic hospitals. The aim of this event is to provide timely updates on the field of pharmacogenomics to local biomedical scientists and health care providers, to inform them on the application of pharmacogenomics in modern medical practice, and to bring together faculty members from universities and research institutes from the local scientific arena working in the field of pharmacogenomics in order to initiate collaborative projects in this field for the benefit of society. Contrary to the Golden Helix Symposia, the Golden Helix Pharmacogenomics Days have a more educational flavor, span between half day to a full day in length, and attract, on average,

Table 9.2 Past Golden Helix Pharmacogenomics Days

| A/A | Year | Date | City | Country |
|-----|------|------|------|---------|
| 1 | 2009 | May 7 | Athens | Greece |
| 2 | 2010 | April 15 | Thessaloniki | Greece |
| 3 | 2011 | April 8 | Alexandroupolis | Greece |
| 4 | | October 7 | Cagliari | Italy |
| 5 | | December 3 | Msida | Malta |
| 6 | 2012 | June 5 | Belgrade | Serbia |
| 7 | | October 20 | Patras | Greece |
| 8 | | November 30 | Brno | Czech Republic |
| 9 | 2013 | May 31 | Zagreb | Croatia |
| 10 | | September 21 | Amsterdam | The Netherlands |
| 11 | | October 10 | Bled | Slovenia |
| 12 | | December 6 | Aviano | Italy |
| 13 | 2014 | May 30 | Sarajevo | Bosnia and Herzegovina |
| 14 | | October 10 | Cardiff | Wales |
| 15 | 2015 | March 6 | Trieste | Italy |
| 16 | | June 22 | Boston, MA | United States of America |
| 17 | 2016 | May 27 | Rotterdam | The Netherlands |
| 18 | | June 16 | Lucca | Italy |
| 19 | | December 1 | Granada[a] | Spain |
| 21 | 2017 | May 4 | Cairo | Egypt |
| 22 | | May 12 | Vienna[a] | Austria |
| 23 | | November 4 | Nicosia | Cyprus |
| 24 | | November 17 | Toulouse[a] | France |

[a]U-PGx Personalized Medicine days (see text for details).

150 participants (mostly between 120 and 220 participants). Registration is free of charge, attracting mostly local participants, but also participants from neighboring countries and regions.

Since early 2009, the Golden Helix Foundation has organized 21 Golden Helix Pharmacogenomics Days in 21 cities in 14 different countries in Europe, Africa, and the United States (Table 9.2; Squassina et al., 2012; Stojiljkovic et al., 2012).

## Golden Helix Summer Schools

Golden Helix Summer Schools are the most recent addition to the Golden Helix conference series. Established as a concept in 2013, and jointly organized under the umbrella of the Golden Helix Foundation and the Genomic Medicine Alliance (see also next chapter), the Golden Helix Summer Schools are international educational activities related to the fields of genomic medicine and biomedical informatics. These educational activities constitute a unique opportunity for researchers around the world to expand their knowledge of the rapidly evolving field of genomic medicine and to exchange innovative ideas in the conductive environment of the Greek islands. Each summer school aims to provide a unique academic program intertwined with

| A/A | Year | Date | Island | Country |
|-----|------|------|--------|---------|
| 1 | 2014 | September 11–15 | Aegina | Greece |
| 2 | 2016 | September 22–26 | Syros | Greece |

Table 9.3 Past Golden Helix Summer Schools

an attractive social program for both invited speakers and participants, while indulging in the Greek summer. Also, the scientific program includes a career development session, in which selected summer school faculty members, coming from the academic, corporate, and/or regulatory arena, engage in lively discussions with the participants, sharing their own experiences from starting their own careers and answering questions, hence providing essential, cutting-edge information on building a successful career in the academic, corporate, and/or regulatory sectors.

In addition to lectures, workshops, course materials, and syllabi, the registration fee, which is kept to a minimum to encourage participation of researchers from developing and lower-income countries, includes accommodation, all meals, and social activities. Emphasis is also given to the social aspect of the Golden Helix Summer Schools, allowing participants to get to know each other in a very informal way, providing them with the chance to visit archaeological sites, to admire the rich history of ancient Greece, and to experience traditional Greek cuisine in the setting of a farewell dinner.

The Golden Helix Summer Schools are organized on a biennial basis; the first event that was organized was in September 2014 on the island of Aegina (Greece), with the theme focused on pharmacogenomics and genomic medicine. The next event was organized in September 2016 on the island of Syros (Greece), with the theme revolving around cancer genomics and treatment individualization (Table 9.3). The next Golden Helix Summer School is planned for September 2018 (26–30 September 2018, again, on the island of Syros) with a tentative theme of genomics of rare diseases, rare cancers, and rare drug outcomes.

## Other Scientific Conferences

Apart from the mainstream Golden Helix conferences outlined above, the Golden Helix Foundation has been involved in the co-organization of other scientific conferences in the field of genome medicine. In particular, the Golden Helix Foundation was engaged in the co-organization, together with local academic entities and in conjunction with the US Food and Drug Administration, of the first and second Latin America Pharmacogenomics Conferences (in San Juan, Puerto Rico, May 10–14, 2010; and in Rio de Janeiro, Brazil, June 28–29, 2012). Similarly, the Golden Helix Foundation participated as local host of the third Global Genomic Medicine Collaborative (G2MC) Conference that was organized in Athens, Greece (April 27–29, 2017), with a theme related to the implementation of genomic medicine in the clinic and will also participate as organising partner of the fourth G2MC Conference that will be organised in Cape Town, South Africa (November 28–30, 2018) with a theme related to implementation of genomic medicine in resource-limited settings.

Also, as of 2016 and as part of its active role in the dissemination and outreaching activities of the Ubiquitous Pharmacogenomics Project (U-PGx; H2020-668353; http://www.upgx.eu), the Golden Helix Foundation has participated in the co-organization, together with local partners of the UPGx Consortium, of the U-PGx Personalized Medicine Days. These events are based exclusively on the concept of Golden Helix Pharmacogenomics Days. There are eight UPGx Personalized Medicine Days planned for the entire duration of the U-PGx project (2016–20), and at the end of the project, a large 4-day symposium, the U-PGx Personalized Medicine Symposium, will be organized in Leiden, the Netherlands, again based exclusively on the concept of Golden Helix Symposia.

## THE GOLDEN HELIX ACADEMY

The Golden Helix Academy is part of the educational activities of the Golden Helix Foundation. It aims to establish, in concert with established academic institutions, e-learning modules related to genomic medicine topics, including but not limited to pharmacogenomics and genomic medicine, public health genomics, genome informatics, economics and health technology assessment in genomics, nutrigenomics, and ethics in genomics (genethics). These e-learning modules lead to academic certificates specifically aimed at the needs of parties from developing countries. These modules are provided in English, as well as in local languages, where applicable.

### Level of Genetic Literacy in the General Public

In general terms, genomic literacy involves two distinct competencies: (1) understanding what genetics and genomics knowledge is and how it is produced, and (2) the capacity to reason and make decisions about socio-scientific issues relevant to genetics and genomics (Roberts, 2007). Studies on the public understanding of genetics and genomics have shown that the general public lacks an accurate understanding of even fairly fundamental genetic concepts, which impacts on its understanding of more complex genetic phenomena (Lanie et al., 2004). Ongoing research suggests that this may also directly impact on decision-making and argumentation about societal issues associated with genomics and its implications for medicine (Lewis and Leach, 2006; Sadler and Fowler, 2006; Sadler and Zeidler, 2005).

There are two important issues that are crucial to the public understanding of genetics. The first is the public perception of science as a process, and of the kind of knowledge it produces, particularly when it comes to the role that genetics and the mass media have played in formulating these perceptions (Kampourakis et al., 2014a). The second issue relates to how people perceive the role of DNA and genes in human health. Genetic determinism appears to be widespread, at least in formal education (but this may not be the case in mass media portrayals of genetics; see Condit et al., 1998; Nelkin and Lindee, 2004). A common view is that there are invariably genes for genetic traits and that single gene defects are responsible even for complex traits. The roots of such misconceptions may be partly found in the way biology is taught in schools, as textbooks often present genetic concepts in a simplistic manner without relating them to the complexities of development (Gericke et al., 2012). Interestingly, even biology teachers may hold simplistic and inaccurate views of genetic determinism (Castera and Clement, 2014). It is therefore unsurprising that pupils may complete their high school education possessing a somewhat naïve, deterministic view of genetics (Mills Shaw et al., 2008).

Given such misconceptions about genetics, it is very challenging to enhance genetic literacy in our society. One important step forward would be to provide the general public (starting even from secondary or even primary school education) with a more accurate portrayal of genetics and genomics (Barnes and Dupré, 2008; Krimsky and Gruber, 2013), and as such, genetics and genomics researchers should be actively involved in education and public communication (Kampourakis et al., 2014a,b; Reydon et al., 2012). Any public communication of genetics should attempt to convey a more accurate view of how genes and the genome as a whole function while emphasizing the complexities of inheritance and development. Furthermore, exploring the possible psychological roots of naïve genetic determinist conceptions will also assist with the development of a more effective genetics/genomics education.

## Conferences to Raise Genetic Awareness in the General Public

From the above it is obvious that it is of the utmost importance to increase genomic awareness in the general public through dedicated conferences, as apart from health care professionals, patients and the general public are equally important stakeholders for the clinical implementation of genomic and personalized medicine.

The Golden Helix Foundation has engaged in the organization of such conferences, and in 2011 it co-organized a conference related to the impact of genetics to society in Athens, Greece (November 4–6, 2011), to enhance awareness of the general public over various issues pertaining to genetics and their application to modern medical practice. This meeting was encouraged by the Howard Hughes Medical Institute. Similarly, in 2012 this conference was repeated jointly with the Eugenides Foundation in Greece (December 2, 2012).

Also, the Golden Helix Foundation organizes satellite events to Golden Helix conferences, specifically aimed at the general public, that deal with issues pertaining to genetics,. Such an event was organized during the 2016 Golden Helix Summer School on September 24, 2016, with a theme evolving around cancer genetics.

Finally, the Golden Helix Foundation, as part of its active involvement in the U-PGx project, plans to organize two major events aimed at increasing the awareness of patients and the general public about the U-PGx project and the usefulness of genome-guided stratification and rationalization of medical treatment modalities. In particular, two major events for patients and the general public will be organized in London (December 2017) and Rome (December 2018) in close cooperation with patients' organizations and advocacy groups so that attendance to these conferences as well as their impact to society is maximized.

## CONCLUSIONS AND FUTURE PERSPECTIVES

In the previous paragraphs, we have summarized the overall educational and outreaching activities of the Golden Helix Foundation, aiming to enrich genomics education among health care professionals and biomedical scientists and at the same time increase genomic literacy among patients and the general public. As already highlighted, the most important feature of the Golden Helix conferences is to ensure that all recent knowledge and developments in the field of genomic and precision medicine are brought to participants from developing countries, who often face financial and logistic constrains to attend similar conference in the United States and Western Europe. At the same time, knowledge and developments in the field are communicated to the delegates by internationally renowned speakers from research centers of excellence. Since November 2008, when the first Golden Helix Symposium was organized in Athens, Greece, there have been 35 Golden Helix conferences that have been organized in 29 major cities in 19 different countries in Europe, the Americas, the Middle East, and Southeast Asia, having attracted more than 6000 participants from more than 100 countries from all five continents. Participants have ranged from students to senior academics and researchers, while several participants have registered to attend recurrent events, indicative of the impact of these scientific events.

Also, on several occasions, and through the partnership with the Genomic Medicine Alliance (Patrinos and Brand, 2014), abstracts from major Golden Helix conferences are published as freely available, as well as PubMed-cited full-text supplements of the international peer-reviewed scientific journal *Public Health Genomics* (Anonymous, 2015, 2016), hence offering Golden Helix conference participants an international forum to maximize exposure of their findings.

Most importantly, the overall environment of these conferences is conductive to establishing partnerships and collaborations among senior scientists, while junior delegates can also benefit by interacting with speakers, which may create possibilities for short-term or even longer term doctoral training or postdoctoral opportunities, which may boost their scientific careers. Several collaborations and research consortia have either formed or strengthened as a result of these conferences, such as in the Balkan and Southeast Asian countries, while various research internships have been offered to junior participants from developing countries by academics and group leaders from major academic centers in Europe and the United States.

These academic activities that foster genomics awareness in patients and the general public and enrich genomics education of clinicians and biomedical scientists, either in the form of the Golden Helix conferences or the Golden

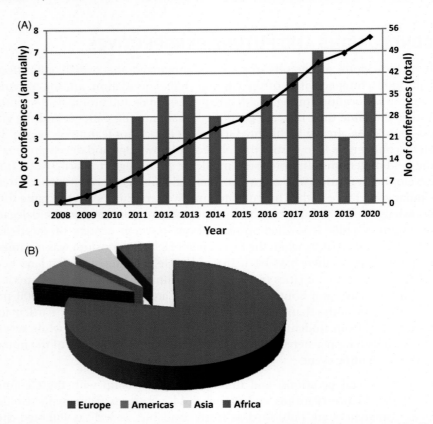

**FIGURE 9.3** (A) The number of Golden Helix conferences and related scientific events co-organized by the Golden Helix Foundation (depicted on an annual basis and in total). The conferences that are planned for 2018–20 are depicted in red. (B) Pie chart representing the geographical distribution of the Golden Helix conferences that have been organized so far, per continent.

Helix Academy, will continue to grow at the same pace (Fig. 9.3) as before. And, when available, in close cooperation with other academic and research networks, such activities always aim to maintain a high standard in genomic medical education and capacity building, specifically addressing the needs of participants from developing countries and low-resource environments.

## Acknowledgments

We are indebted to the various academic and corporate entities and charitable organizations worldwide that have supported the various Golden Helix conferences and to the numerous organizations that have endorsed these events. Also, we thank the administrative and management team of the Golden Helix Foundation for their valuable administrative support.

## References

Anonymous 2015, 2015. Golden Helix Symposium—next generation pharmacogenomics. March 11–13, 2015, Kuala Lumpur. Malaysia: abstracts. Public Health Genomics 18 (Suppl. 1), 1–51.

Anonymous, 2016, 2016. 2016 Golden Helix Summer School—Cancer Genomics and Individualized Therapy. 22–26 September 2016, Syros Island, Greece: Abstracts. Public Health Genomics 19 (Suppl. 1), 1–11.

Barnes, B., Dupré, J., 2008. Genomes and What to Make of Them. University of Chicago Press.

Castera, J., Clement, P., 2014. Teachers' conceptions about genetic determinism of human behaviour: a survey in 23 countries. Sci. Edu. 23, 417–443.

Condit, C.M., Ofulue, N., Sheedy, K.M., 1998. Determinism and mass-media portrayals of genetics. Am. J. Hum. Genet. 62, 979–984.

Fortina, P., Al Khaja, N., Al Ali, M.T., Hamzeh, A.R., Nair, P., Innocenti, F., Patrinos, G.P., Kricka, L.J., 2014. Genomics into healthcare: the 5th Pan Arab human genetics conference and 2013 Golden Helix Symposium. Hum. Mutat. 35, 637–640.

Gericke, N., Hagberg, M., Carvalho Santos, V., Joaquim, L.M., El-Hani, C., 2012. Conceptual variation or incoherence? Textbook discourse on genes in six countries. Sci. Edu. 23, 381–416.

Haga, S.B., O'Daniel, J.M., Tindall, G.M., Mills, R., Lipkus, I.M., Agans, R., 2012a. Survey of genetic counselors and clinical geneticists' use and attitudes toward pharmacogenetic testing. Clin. Genet. 82, 115–120.

Haga, S., Burke, W., Ginsburg, G., Mills, R., Agans, R., 2012b. Primary care physicians' knowledge and experience with pharmacogenomics testing. Clin. Genet. 82, 388–394.

Kampourakis, K., Reydon, T.A., Patrinos, G.P., Strasser, B.J., 2014a. Genetics and society—educating scientifically literate citizens: introduction to the thematic issue. Sci. Edu. 23, 251–258.

Kampourakis, K., Vayena, E., Mitropoulou, C., van Schaik, R.H., Cooper, D.N., Borg, J., Patrinos, G.P., 2014b. Key challenges for next-generation pharmacogenomics. EMBO Rep. 15, 472–476.

Kricka, L.J., Di Resta, C., 2013. Translating genes into health. Nat. Genet. 45, 4–5.

Krimsky, S., Gruber, J. (Eds.), 2013. Genetic Explanations: Sense and Nonsense. Harvard University Press, Cambridge MA.

Lanie, A.D., Jayaratne, T.E., Sheldon, J.P., Kardia, S.L., 2004. Exploring the public understanding of basic genetic concepts. J. Genet. Couns. 13, 305–320.

Le Caignec, C., Redon, R., 2009. Copy number variation goes clinical. Genome Biol. 10, 301.

Lewis, J., Leach, J., 2006. Discussion of socio-scientific issues: the role of science knowledge. Int. J. Sci. Edu. 28, 1267–1287.

Mai, Y., Mitropoulou, C., Papadopoulou, X.E., Vozikis, A., Cooper, D.N., van Schaik, R.H., Patrinos, G.P., 2014. Critical appraisal of the views of healthcare professionals with respect to pharmacogenomics and personalized medicine in Greece. Per. Med. 11, 15–26.

Mills Shaw, K.R., van Horne, K., Zhang, H., Boughman, J., 2008. Essay contest reveals misconceptions of high school students in genetics content. Genetics 178, 1157–1168.

Nelkin, D., Lindee, S.M., 2004. The DNA Mystique: The Gene as a Cultural Icon. University of Michigan Press, Ann Arbor, MI.

Patrinos, G.P., Brand, A., 2014. Public health genomics joins forces with the genomic medicine alliance. Public Health Genomics 17, 125–126.

Patrinos, G.P., Innocenti, F., 2010. Pharmacogenomics: paving the path to personalized medicine. Pharmacogenomics 11, 141–146.

Patrinos, G.P., Petersen, M.B., 2009. Copy number variation and genomic alterations in health and disease. Genome Med. 1, 21.

Patrinos, G.P., Innocenti, F., Cox, N., Fortina, P., 2011. Genetic analysis in translational medicine: the 2010 Golden Helix Symposium. Hum. Mutat. 32, 698–703.

Patrinos, G.P., Baker, D.J., Al-Mulla, F., Vasiliou, V., Cooper, D.N., 2013. Genetic tests obtainable through pharmacies: the good, the bad, and the ugly. Hum. Genomics 7, 17.

Reydon, T.A., Kampourakis, K., Patrinos, G.P., 2012. Genetics, genomics and society: the responsibilities of scientists for science communication and education. Per. Med. 9, 633–643.

Roberts, D.A., 2007. Scientific literacy/science literacy. In: Abell, S.K., Lederman, N.G. (Eds.), Handbook of Research on Science Education. Erlbaum, Mahwah, NJ, pp. 729–780.

Sadler, T.D., Fowler, S.R., 2006. A threshold model of content knowledge transfer for socioscientific argumentation. Sci. Edu. 90, 986–1004.

Sadler, T.D., Zeidler, D.L., 2005. The significance of content knowledge for informal reasoning regarding socioscientific issues: applying genetics knowledge to genetic engineering issues. Sci. Edu. 89, 71–93.

Squassina, A., Manchia, M., Manolopoulos, V.G., Artac, M., Lappa-Manakou, C., Karkabouna, S., Mitropoulos, K., Del Zompo, M., Patrinos, G.P., 2010. Realities and expectations of pharmacogenomics and personalized medicine: impact of translating genetic knowledge into clinical practice. Pharmacogenomics 11, 1149–1167.

Squassina, A., Severino, G., Grech, G., Fenech, A., Borg, J., Patrinos, G.P., 2012. Golden Helix Pharmacogenomics Days: educational activities on pharmacogenomics and personalized medicine. Pharmacogenomics 13, 525–528.

Stanek, E.J., Sanders, C.L., Taber, K.A., Khalid, M., Patel, A., Verbrugge, R.R., Agatep, B.C., Aubert, R.E., Epstein, R.S., Frueh, F.W., 2012. Adoption of pharmacogenomic testing by US physicians: results of a nationwide survey. Clin. Pharmacol. Ther. 91, 450–458.

Stojiljkovic, M., Fazlagic, A., Dokmanovic-Krivokapic, L., Nikcevic, G., Patrinos, G.P., Pavlovic, S., Zukic, B., 2012. 6th Golden Helix Pharmacogenomics Day: pharmacogenomics and individualized therapy. Hum. Genomics 6, 19.

# CHAPTER 10

# The Genomic Medicine Alliance: A Global Effort to Facilitate the Introduction of Genomics into Healthcare in Developing Nations

David N. Cooper*, Christina Mitropoulou**, Angela Brand[†],
Vita Dolzan[‡], Paolo Fortina[§], Federico Innocenti[¶], Ming T.M. Lee[††,‡‡],
Milan Macek Jr.[¶¶], Konstantinos Mitropoulos**, Fahd Al-Mulla[§§],
Barbara Prainsack***, Ron H. van Schaik[†††], Alessio Squassina[‡‡‡],
Domenica Taruscio[¶¶¶], Effy Vayena[§§§], Athanassios Vozikis****,
Marc S. Williams[††], Bauke Ylstra[††††], George P. Patrinos[‡‡‡‡,§§§§§]

*Institute of Medical Genetics, Cardiff University, Cardiff, United Kingdom;
**The Golden Helix Foundation, London, United Kingdom; [†]University of Maastricht,
Institute of Public Health Genomics, Maastricht, The Netherlands; [‡]University of
Ljubljana, Ljubljana, Slovenia; [§]Thomas Jefferson University, Kimmel Cancer Center,
Philadelphia, PA, United States; [¶]Institute of Pharmacogenomics and Individualized
Therapy, University of North Carolina, Chapel Hill, NC, United States; [††]Genomic Medicine
Institute, Danville, PA, United States; [‡‡]RIKEN Center for Integrative Medical Sciences,
Yokohama, Japan; [¶¶]Charles University Prague, Institute of Biology and Medical Genetics,
Prague, Czech Republic; [§§]Kuwait University, Health Sciences Center, Kuwait City,
Kuwait; ***King's College London, London, United Kingdom; [†††]Erasmus University
Medical Center, Rotterdam, The Netherlands; [‡‡‡]University of Cagliari, Cagliari,
Italy; [¶¶¶]National Centre for Rare Diseases, Istituto Superiore di Sanità, Rome, Italy;
[§§§]University of Zurich, Epidemiology Biostatistics and Prevention Institute, Zurich,
Switzerland; ****University of Piraeus, Piraeus, Greece; [††††]VU University Medical Center,
Amsterdam, The Netherlands; [‡‡‡‡]University of Patras School of Health Sciences, Patras,
Greece; [§§§§§]United Arab Emirates University, Al-Ain, United Arab Emirates

## INTRODUCTION

Genomic medicine aims to utilize the individual's genomic information to support the clinical decision-making process (Manolio et al., 2015). In recent years, significant advances have been made in understanding the molecular basis of a wide range of human inherited diseases and cancers with the potential to improve disease prognosis and treatment (Kilpinen and Barrett, 2013).

**173**

Genomic Medicine in Emerging Economies. http://dx.doi.org/10.1016/B978-0-12-811531-2.00010-2

At the same time, genomic technology has progressed rapidly, with a variety of new high-throughput genome-wide screening and massively parallel sequencing approaches becoming available (Gullapalli et al., 2012). As a result, genomic information is becoming more readily available with the potential to play a role in diagnosis, disease risk-stratification, medication selection and dosing, carrier screening, and other emerging uses. This constitutes the basis of *genomic medicine*, a relatively new discipline that aims to enhance opportunities for disease prevention and the customization of patient care, including the personalization of conventional and new therapeutic interventions (Lazaridis et al., 2014).

Genomic medicine is closely linked to the concept of or *personalized medicine*, which refers to the aim to tailor diagnosis and treatment more closely to the individual characteristics of patients (European Science Foundation, 2012). Although personalized medicine as a concept has gained particular currency within the last two decades, its central concept was proposed around 400 BC; Hippocrates of Kos (460–370 BC) stated that "… it is more important to know what kind of person suffers from a disease than to know the disease a person suffers." The first application of genomic medicine can be said to have been codified in the Talmud (Yevamot 64b), where it is stated [Rabbi Judah the Prince's ruling (2nd century BC)] that if a woman's first two children had died from blood loss after circumcision, the third son should be exempted from circumcision. Rabbi Simeon ben Gamliel disagreed and ruled that the third son might be circumcised, but if this infant also died then the fourth child should not be circumcised. These ancient examples could be seen to encapsulate the essence of personalized medicine when people's personal circumstances and characteristics mean that not everybody is treated the same. Today, genomics yields important information for personalization. For this reason, several international organizations and research consortia have been formed with the stated goal of supporting the translation of genomic research into clinical practice so that genomic medicine can ultimately be used to benefit the global community.

The Genomic Medicine Alliance (GMA; http://www.genomicmedicinealliance. org/; Cooper et al., 2014) is a newly established global academic research network, which aims to build and strengthen collaborative ties between academics, researchers, regulators, and those members of the general public interested in genomic medicine. The GMA activities are focusing in particular on developing countries and low-resource environments. Herein, we use the notion "developing country" as an environment where (1) resources assigned for genomics research are scarce, (2) access to genomics knowledge and information is low, (3) genomic implementation is limited, (4) genomics education is relatively poor, and/or (5) collaborative opportunities in genomics research with other institutions are rare for a number of reasons, such as geographical, societal, economical, and/or political (see also Chapter 1).

## MISSION AND GOALS OF THE GENOMIC MEDICINE ALLIANCE

The GMA aims to do the following: (1) encourage and catalyze multidisciplinary collaborative research between partner institutions and/or scientists, with at least one partner affiliated with an institute in a developing country; (2) liaise between research organizations, clinical entities, and regulatory agencies on topics related to genomic medicine; (3) facilitate the introduction of pharmacogenomics and advanced'omics technologies into mainstream clinical practice; (4) propose international guidelines and draw up recommendations for activities pertaining to genomic medicine, in close collaboration with other scientific academic entities, agencies, and regulatory bodies; and (5) develop independently and coordinate, in close collaboration with partner institutions, educational activities in the sphere of genomic medicine.

The GMA aims to foster collaboration in genomics research between developed and developing/low-resourced countries, seeking to ensure that such collaboration is beneficial to all parties concerned. Developing countries should benefit from training opportunities, knowledge exchange, and the expansion of transnational networks, whereas developed countries could expect to benefit through comparative work on ethnically diverse populations that have not yet been well studied, and by having access to families with rare diseases or unique clinical features, especially where the developing countries are characterized by a higher incidence of consanguinity and/or well-defined founder populations (see also below); this includes patients coming from ethnic communities within developed countries that are characterized by a high rate of consanguineous marriage. Considering that approximately 85% of the world's population live in developing countries, this represents a major challenge to access and engage a hitherto neglected group of individuals with rare diseases (Cooper et al., 2014). Rare diseases are not only important in terms of improving our understanding of the pathology to benefit the affected patients and their families, but they have the potential to provide key insights that can lead to a better understanding of gene function in both health and disease (Collins, 2011).

A key aim of GMA is paving the way from genomics research to genomic medicine by encouraging and undertaking multicenter research projects in

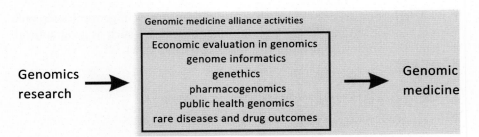

**FIGURE 10.1** Graphical depiction of the Genomic Medicine Alliance research activities that aim to translate genomics research and pharmacogenomics into genomic medicine (see text for details). Research disciplines are listed in alphabetical order (this order does not imply any prioritization of research activity).

key subdisciplines. To this end, GMA activities aim to contribute to the transition from genomics and pharmacogenomics research to genomic medicine, viz., public health genomics, ethics in genomics (or "genethics"), genome informatics, the genetics education of health care professionals, the genetics awareness of the general public, and health economic evaluation in relation to genomic medicine. This has been previously represented pictorially as an ancient Greek temple, where genomics and pharmacogenomics research represents the bedrock of genomic medicine, and where the various subdisciplines are held above the supporting pillars that must be carefully erected for the superstructure of genomic medicine to hold (Fig. 10.1). At present, although the foundations of genomic medicine are becoming stronger and being ascribed ever-increasing hopes and expectations, the pillars themselves are still largely under construction.

GMA research activities are supervised by an international scientific advisory committee comprising 16 internationally renowned scientists in the field from all over the world (Table 10.1). Administrative assistance is provided by the Golden Helix Foundation (http://www.goldenhelix.org/) staff. Registration with the GMA is free of charge in order to encourage the participation of researchers from developing/low-income countries and emerging economies. Upon registration, members specify their research interests so that they can be directed to research projects and training opportunities that suit their specific needs.

The GMA has recently established the concept of "GMA Ambassadors" within the network, aiming to actively engage dynamic, mostly early-career but also senior scientists with a keen interest in genomic medicine, who are interested in expanding the GMA network in their own territory. In particular, the role of GMA Ambassadors will be to (1) increase awareness of GMA activities and events among their peers and colleagues, (2) attract new members to the GMA

Table 10.1 Members of the Genomic Medicine Alliance International Scientific Advisory Committee (in Alphabetical Order by Continent)

| Continent | Number | Member | Country |
|---|---|---|---|
| Asia/Middle East | 1 | Fahd Al-Mulla | Kuwait |
| | 2 | Ming Ta Michael Lee | Japan |
| Americas | 3 | Paolo Fortina | USA |
| | 4 | Federico Innocenti | USA |
| | 5 | Marc S. Williams | USA |
| Europe | 6 | Angela Brand | The Netherlands |
| | 7 | David N. Cooper | United Kingdom |
| | 8 | Vita Dolzan | Slovenia |
| | 9 | Milan Macek Jr | Czech Republic |
| | 10 | George P. Patrinos | Greece |
| | 11 | Barbara Prainsack | United Kingdom |
| | 12 | Ron H. van Schaik | The Netherlands |
| | 13 | Alessio Squassina | Italy |
| | 14 | Domenica Taruscio | Italy |
| | 15 | Effy Vayena | Switzerland |
| | 16 | Athanassios Vozikis | Greece |
| | 17 | Bauke Ylstra | The Netherlands |

through social media and other means, (3) contribute and/or comment on articles posted on the GMA portal pertaining to their area of expertise and territories, and (4) represent the GMA, if required, at scientific events and conferences that the GMA Ambassadors attend.

GMA research activities span four different working groups: genome informatics, genomics and pharmacogenomics, public health genomics, and health economics. Each of the working groups' activities are coordinated by the corresponding Working Group and Activity leaders in conjunction with Senior National Representatives from each of the >70 countries from which the >1300 current GMA members (January 2018) originate. Some of the key GMA research projects are outlined below.

## CURRENT RESEARCH PROJECTS AMONG GENOMIC MEDICINE ALLIANCE MEMBERS

### Genomics and Pharmacogenomics Working Group

Pharmacogenomics aims to rationalize drug treatment by optimizing the balance between treatment efficacy and toxicity based on a comprehensive understanding of the impact of genomic variants on drug metabolism combined with other patient-based and environmental factors. The GMA Pharmacogenomics Working Group, in close collaboration with the Golden Helix Foundation, is currently taking part in the Euro-PGx project (http://

www.goldenhelix.org/index.php/research/pharmacogenomics-in-europe) in which 26 European populations are participating. More specifically, the Euro-PGx project aims to (1) determine the population-specific allele frequencies of pharmacogenomics variants to optimize medication choice and dose and minimize adverse reactions by genotyping 1936 pharmacogenomically relevant genetic variants in 231 absorption, distribution, metabolism, excretion, and toxicity (ADMET)-related pharmacogenes, which would assist in prioritizing medication selection in participating developing countries; and (2) develop off-the-shelf solutions for pharmacogenomic testing in participating developing countries. There are significant interpopulation pharmacogenomic allele frequency differences, particularly in seven clinically actionable pharmacogenes in seven European populations that affect the drug efficacy and/or toxicity of 51 medication treatment modalities. This includes differences observed in the prevalence of high-risk genotypes in these populations in the *CYP2D6*, *CYP2C9*, *CYP2C19*, *CYP3A5*, *VKORC1*, *SLCO1B1*, and *TPMT* pharmacogenes, resulting in notable differences in drug response, such as the genotype-based warfarin dosing between these populations (Mizzi et al., 2016). These findings can be used not only to develop guidelines for medication prioritization, but most importantly to facilitate the integration of pharmacogenomics and to support preemptive pharmacogenomic testing. Replication of these findings in larger population samples would permit the establishment of a rational framework for pharmacogenomic testing in developing countries to support the incorporation of country-specific population characteristics in a standardized fashion.

At the same time, the GMA Pharmacogenomics Working Group has sought to provide proof-of-principle of the use of whole-genome sequencing for pharmacogenomic testing, by resequencing with high coverage almost 500 whole genomes, mostly from Caucasian populations. This project not only revealed a vast number of novel potentially functional variants in a total of 231 pharmacogenes, as indicated by *in silico* analysis, but has also demonstrated the value of whole-genome sequencing for pharmacogenomic testing by capturing over 18,000 variants in these pharmacogenes, in contrast to just over 250 variants that would have been identified in these genes by using the most comprehensive pharmacogenomics assay currently available (Mizzi et al., 2014).

Identification of genomic variants and structural alterations that guide therapy selection for patients with cancer has nowadays become routine in many clinical centers. The majority of genomic assays used for solid tumor profiling employ next-generation resequencing to interrogate mostly somatic but also germline variants because they can be more easily identified and interpreted.

## Genome Informatics Working Group

Documentation of the incidence of genetic disorders in different populations, particularly in those developing countries with a high incidence of genetic diseases and/or consanguinity, can be particularly helpful in the context of adopting national prevention and screening programs (Patrinos, 2006). GMA members have actively participated in the development of new or the update of existing national/ethnic genetic databases for several populations in GMA member territories, such as Greece, Serbia, Kuwait, Egypt, and Tunisia, by using the newly upgraded ETHNOS software (Viennas et al., 2017).

The result is that the ETHNOS software supports, in its present format, the development of a large number of national genetic databases (Papadopoulos et al., 2014) based on the data warehouse principle and preexisting guidelines (Patrinos et al., 2011). These databases are being assigned to senior human geneticists in the corresponding populations in order to coordinate their curation and stimulate data enrichment and expansion.

Also, recognizing the need to provide a comprehensive resource for pharmacogenomics biomarker information for those drugs that are approved by regulatory authorities, members of the Genome Informatics Working Group have extracted from the published literature all pharmacogenomic biomarkers that relate to the FDA- and EMA-approved drugs with pharmacogenomic information in their label and made them available in a database that triangulates between drugs, genes, and pharmacogenomics biomarkers. The DruGeVar database (Dalabira et al., 2014) was developed jointly with the Global Genomic Medicine Collaborative Pharmacogenomics Working Group and contains a total of 545 records, involving correlations between 91 drugs, 13 genomic loci and 98 genomic variants, previously implicated in variable drug responses, both in terms of efficacy and toxicity. All DruGeVar database records depict drug/gene combinations that have been approved by any or both of the two major regulatory agencies and are made freely available to the public for data querying. Recently, the DruGeVar database was included as a plug-in module for the pharmacogenomic biomarkers module of FINDbase database (Viennas et al., 2017). Overall, being one of the few electronic resources in the field of pharmacogenomics, the DruGeVar database is expected to contribute toward bridging the gap between pharmacogenomics research findings and clinical practice.

Owing to the rapid evolution of next-generation sequencing, the past decade has seen the characterization of both somatic and/or germline alterations in a wide range of cancers, generating a large body of information pertaining to how cancer develops, evolves, and reacts to various treatment modalities (Macintyre et al., 2016). Also, a considerable number of genomic variants have previously been reported to be causative of, or associated with, tumor progression or an increased risk for various types of cancer (Hanahan and Weinberg, 2011). Members of this Working Group aim to define some of the key tools for genomic testing that primary care practitioners and specialists should know about when considering how to treat cancer patients.

At the same time, members of this Working Group have undertaken a study to identify cancer predisposition (germline) variants in apparently healthy individuals with no cancer history in the family by using a next-generation sequencing strategy. Such an approach aimed to identify genomic, particularly novel, variants that might predispose to various types of cancer so that such information could help in the assessment of personalized cancer-susceptibility risk from genome sequence data (Karageorgos et al., 2015). This includes both genomic variants in genes (like, e.g., *BRCA1/2* genes) that have previously been associated with heritable risk conditions, and also risk alleles that are known to increase cancer predisposition risk, but not in a Mendelian sense. Indeed, a small fraction (3%) of a large number of variants (571 variants in total) previously associated with cancer predisposition has been shown to be potentially pathogenic in the members of two families. This approach could be adopted for other types of complex genetic disorder in order to identify variants of potential pathological significance.

## Public Health Genomics Working Group

Public Health Genomics represents the responsible and effective translation of genome-based knowledge and technologies into public policy and health services for the benefit of population health (Burke et al., 2006). The GMA Public Health Genomics Working Group is undertaking national and transnational studies to improve our understanding of the level of public awareness of genetics, including their attitudes to genomic testing and the level of genetics education of health care professionals (i.e., physicians, pharmacists, etc.). So far, such surveys have yielded some interesting findings (Mai et al., 2014), highlighting the relative lack of genetics education of health care professionals and genetic awareness and literacy of the general public as perhaps some of the biggest obstacles to the widespread implementation of genomic medicine (Kampourakis et al., 2014). A detailed stakeholder analysis that aimed to comprehend the attitudes and map the genomic medicine policy environment was undertaken, serving as a database for assessments of the policy's content, the major players, their power and policy positions, their interests and networks, and coalitions that interconnect them (Mitropoulou et al., 2014). These findings should contribute to the selection and implementation of policy measures that will expedite the adoption of genomics into conventional medical interventions; such studies are currently being replicated in other countries, under the umbrella of the GMA.

These studies have also shown the general utility of genomic testing for individuals, including the public's remarkable level of interest in participating in genomic research (Reydon et al., 2012; Demmer and Waggoner, 2014). Such surveys have already been replicated in other European countries (Pisanu et al., 2014) and are currently being conducted in Southeast Asia and the Middle East, partly supported by the Golden Helix Foundation, thereby confirming initial findings and highlighting the need to harmonize genomics education and to raise genomics awareness among the general public. To this end, GMA members co-organize educational events revolving around pharmacogenomics and genomic medicine in various European countries; these are endorsed by the GMA and partly funded by the Golden Helix Foundation and other entities (Squassina et al., 2012).

Several ethical issues confront those who are committed to the practice of genomic medicine, including the regulation of genetic testing, the governance of genetic research, and genomic data sharing in an ethical and publicly accountable way (Kampourakis et al., 2014). This Working Group also explores the landscape of direct-to-consumer (DTC), beyond the clinic (Prainsack and Vayena, 2013), and over-the-counter (OTC) genetic tests in various European countries, including Greece, Slovenia, Italy, and Serbia. From these undertakings, it is particularly important to harmonize policies that safeguard the general public and ensure that they become better informed with respect to the various attendant risks from this type of testing. Currently, regulation of these issues is lacking in many European countries, as well as at a central level in the form of a directive of the European Medicines Agency for both OTC and DTC genetic testing (Kricka et al., 2011). The GMA has recently produced an opinion article to highlight the various types of OTC genetic tests currently available (Patrinos et al., 2013). GMA members are also working in close cooperation with national genetic societies and national ethics committees to establish guidelines to cover ethical, legal, and social issues pertaining to genetic testing.

Furthermore, in an effort to resolve the ambiguity regarding the utility of nutrigenomics testing given our current level of knowledge, the GMA Public Health Genomics Working Group has encouraged the meta-analysis of a number of studies related to 38 genes included in nutrigenomics tests provided by various private genetic testing laboratories, aiming to identify possible associations between the genes of interest and dietary intake and/or nutrient-related pathologies. No specific and statistically significant association was observed for any of the 38 genes, whereas in those cases in which a weak association was demonstrated, evidence was based on a limited number of studies (Pavlidis et al., 2015). This study has demonstrated that although nutrigenomics research is a promising area for genomic investigation, solid scientific evidence is currently lacking, and as such, commercially available nutrigenomics tests cannot be recommended. This is consistent with the 2014 position statement from the Academy of Nutrition and Dietetics, indicating that "It is the position of the Academy of Nutrition and Dietetics that nutritional genomics provides insight into how diet and genotype interactions affect phenotype" (Camp and Trujillo, 2014). On the contrary, it has been suggested that assessment and synthesis of nutrigenomics data should be carried out on an ongoing basis at periodic intervals and/or when there is a specific demand for a synthesis of the available evidence, and, importantly, in ways that are transparent where potential conflict of interests are fully disclosed by the parties involved (Pavlidis et al., 2016).

## Health Economics Working Group

A key factor in expediting the adoption of genomic medicine in clinical practice would be the demonstration of its cost-effectiveness (as the "fourth hurdle" in health care, after safety, efficacy, and quality). The real cost-effectiveness of involving genomics in medicine is as yet unknown apart from some rather limited studies in pharmacogenomics and hereditary cancer syndromes. Although it is vital to perform cost-effectiveness analyses for the implementation of genomic medicine in developing countries, there are only a handful of such studies reported in the literature (Snyder et al., 2014). Realizing cost-effectiveness would be a crucial step toward convincing policy makers of the utility of genomics in health care as a means to reduce the overall treatment costs, as well as to reduce the overall burden and minimize consequences of disease at the national level (Payne and Shabaruddin, 2010). Currently, the GMA Health Economics Working Group has been successfully engaged in assessing the cost-effectiveness of genome-guided treatment modalities in developing countries. In particular, GMA members have participated in a prospective study to assess the cost-effectiveness of genome-guided warfarin treatment in Croatia, where it has been shown that genome-guided warfarin treatment may represent a cost-effective therapy option for the management of elderly patients with atrial fibrillation who developed ischemic stroke in Croatia, with an estimated incremental cost-effectiveness ratio of the pharmacogenomics-guided versus the control groups of €31,225/quality-adjusted life year (Mitropoulou et al., 2015). Also, a retrospective economic analysis of genome-guided clopidogrel treatment in Serbia indicated that pharmacogenomics-guided clopidogrel treatment may represent a cost-saving approach for the management of myocardial infarction patients undergoing primary percutaneous coronary intervention in Serbia (Mitropoulou et al., 2016).

Members of the GMA Health Economics Working Group have developed and evaluated standardized methodologies for the economic evaluation of genomic medicine (Fragoulakis et al., 2016, 2017), which, in addition to the already existing battery of economic evaluation models in genomic medicine (Annemans et al., 2013), will be of the utmost importance as innovative tools for performing cost-effectiveness analyses in such a rapidly evolving discipline. To this end, the GMA has endorsed the production of a related textbook, coauthored by two GMA Scientific Advisory Committee members, published by Elsevier/Academic Press in early 2015 (Fragoulakis et al., 2015).

Lastly, the issue of pricing and reimbursement was the topic for the GMA Health Economics Working Group, given the lack of harmonization between pricing and reimbursement policies between European countries, contrary to the situation pertaining in the United States (Logue, 2003).

As a first step, the general strategy toward pricing and reimbursement for genomic medicine in Europe has been outlined, providing an overview of the rationale and basic principles guiding the governance of genomic testing services, clarifying their objectives, and allocating and defining responsibilities among stakeholders, focusing on different EU countries' health care systems. Particular attention was paid to issues pertaining to pricing and reimbursement policies, the availability of essential genomic tests, differing between various countries owing to differences in disease prevalence and public health relevance, the prescribing and use of genomic testing services according to existing or new guidelines, budgetary and fiscal control, the balance between price and access to innovative testing, monitoring and evaluation for cost-effectiveness and safety, and the development of research capacity (Vozikis et al., 2016).

Subsequently, it is hoped that a more technical analysis would lead to a robust policy in relation to pricing and reimbursement in genomic medicine, thereby contributing to an effective and sustainable health care system that will prove beneficial to the economy at large.

## ESTABLISHMENT OF THE DRIFT CONSORTIUM

Consistent with one of the stated goals of the GMA, which is the fruitful engagement between research groups from developing and developed countries to study families with rare diseases or unique clinical features (especially countries with a higher incidence of consanguinity and/or well-defined founder populations), the GMA participated in the establishment of the Discovery Research Investigating Founder Population Traits (DRIFT) Consortium.

In early 2016, a call for research collaboration was made by the Regeneron Genetics Center (RGC) and the GMA, aimed specifically at developing countries. DRIFT aims to understand the genetic architecture of founder populations throughout the world with direct impact on human health and disease. The DRIFT Consortium aims to catalog population-specific allelic architecture, to understanding the biological and functional consequences of specific genomic variants identified, and to share and establish best-practice approaches to relieve disease burden in these populations. DRIFT is planning two tiers of collaboration models:

1. Tier 1 aims to canvas the allelic architecture of the population by exome sequencing and DNA microarrays from relatively unrelated individuals. Several hundred de-identified samples will be analyzed at the RGC that will provide high-depth exome sequence and genome-wide association data, derived from DNA microarrays, to be returned to the collaborator free of charge. There would be no need to exchange phenotype information, and if the joint sequence data were used for any genotype–phenotype analyses, the results would be shared with RGC. Most importantly, the collaborator is free to publish results derived from this effort.

2. Tier 2 aims to establish a collaborative effort focused on novel gene discovery for phenotypes of mutual interest. In this tier, an academic collaboration model is established, in which the collaborator and the RGC jointly develop the research plan. In this tier, a much larger number of DNA samples (100s–10,000s) is provided by the collaborator, and again, like Tier 1, RGC provides all exome sequence data to the collaborator free of charge. De-identified phenotype data is shared, data analyses of the combined sequence and phenotype data set are performed collaboratively, and each party is free to use the data set for its own internal research. Again, collaborators are encouraged to publish results, and each party is free to use published results for any and all purposes.

A short-form material transfer agreement is used to govern the collaboration in both tiers. For both models, data and results will be broadly shared with the research community, and if exciting new results are generated from a Tier 1 or Tier 2 collaboration, there will be the potential for the design and funding of follow-up "genotype-first call-back" studies for additional collaborative research to delve more deeply into biological mechanisms and pathways. Such an approach is expected to attract institutions and research groups from developing countries in Europe, Latin America, and the Middle East, which have founder populations bearing some very important features readily available for analysis.

## CONCLUSIONS AND FUTURE PERSPECTIVES

The GMA is a new initiative in the field of genomic medicine, with the primary goal to develop a network focusing on the translation of genomic knowledge into clinical use, with a special focus on the participation of developing countries and of low-resource settings.

The GMA has several unique features as a research network, in which it differs from existing consortia and initiatives in this field (Manolio et al., 2015). First, membership is free of charge, which is important to attract members from developing countries. Second, it has a flat governance structure, comprising the scientific advisory and the steering committees. Third, this network has a stated goal and commitment to bring together genomics research institutions from developing countries with those from developed countries (Cooper et al., 2014).

Ever since its establishment, the expansion of the GMA membership base has progressed at a very rapid pace, currently consisting of over 1300 members from >70 countries worldwide, from academia as well as from corporate and regulatory sectors, including developing countries in the Middle East, Asia, and Latin America.

In 2014 an important milestone for GMA was the agreement with Karger to establish the international peer-reviewed journal, *Public Health Genomics* (http://www.karger.com/Journal/Home/224224) as the Official GMA journal (Patrinos and Brand, 2014). *Public Health Genomics* is the leading bimonthly international journal, published by Karger (Editor-In-Chief: Nicole Probst-Hensch) and focusing on the translation of genome-based knowledge and technologies into public health, health policies, and health care as a whole. This partnership provides GMA members not only with a highly respected forum to publish their original research findings but also with discounts on the journal's annual subscription, open access fees, and Karger books.

In addition, and in order to support the transnational mobility of students and junior researchers, the GMA plans to launch short- and long-term research fellowships for early-stage researchers from developing countries to pursue research in centers of excellence in developed countries. The GMA envisages doing this in collaboration with the Golden Helix Foundation and other charities. Last but not least, the GMA will continue to endorse conferences and educational activities in the field of genomic medicine in Europe, the Middle East, Latin America, and Southeast Asia. Indeed, since 2014 the GMA has established, in conjunction with the Golden Helix Foundation, the Golden Helix Summer Schools (http://summerschools.goldenhelix.org/; see also Chapter 4). This international initiative in the field of genomic medicine and genome informatics aims to provide researchers around the world with the opportunity to expand their knowledge in these rapidly evolving disciplines.

In essence, the GMA aspires to become a focal point for harmonizing research activities in the field of genomic medicine between developed and developing countries while helping to pave the way for a smoother transition from genomics research to genomic medicine.

## Acknowledgments

The authors acknowledge the invaluable assistance of the GMA Working Groups and activity leaders, and of the Senior National Representatives and GMA Ambassadors Theodora Katsila, Maja Stojiljkovic, Ioanna Maroulakou, and Rossana Roncato in relation to coordinating GMA activities and expanding its membership basis.

## References

Annemans, L., Redekop, K., Payne, K., 2013. Current methodological issues in the economic assessment of personalized medicine. Value Health 16, S20–S26.

Burke, W., Khoury, M.J., Stewart, A., Zimmern, R.L., 2006. The path from genome-based research to population health: development of an international public health genomics network. Genet. Med. 8, 451–458.

Camp, K.M., Trujillo, E., 2014. Position of the Academy of Nutrition and Dietetics: nutritional genomics. J. Acad. Nutr. Diet. 114, 299–312.

Collins, F.S., 2011. The promise and payoff of rare diseases research, NIH Medline Plus. http://www.nlm.nih.gov/medlineplus/magazine/issues/spring11/articles/spring11pg2-3.html

Cooper, D.N., Brand, A., Dolzan, V., Fortina, P., Innocenti, F., Lee, M.T., Macek, M., Al-Mulla, F., Prainsack, B., Squassina, A., Vayena, E., Vozikis, A., Williams, M.S., Patrinos, G.P., 2014. Bridging genomics research between developed and developing countries: the Genomic Medicine Alliance. Pers. Med. 11, 615–623.

Dalabira, E., Viennas, E., Daki, E., Komianou, A., Bartsakoulia, M., Poulas, K., Katsila, T., Tzimas, G., Patrinos, G.P., 2014. DruGeVar: an online resource triangulating drugs with genes and genomic biomarkers for clinical pharmacogenomics. Public Health Genomics 17, 265–271.

Demmer, L.A., Waggoner, D.J., 2014. Professional medical education and genomics. Annu. Rev. Genomics Hum. Genet. 15, 507–516.

European Science Foundation (ESF), 2012. Personalised medicine for the European citizen—towards more precise medicine for the diagnosis, treatment and prevention of disease. Strasbourg ESF, Available from: http://archives.esf.org/fileadmin/Public_documents/Publications/Personalised_Medicine.pdf

Fragoulakis, V., Mitropoulou, C., Williams, M.S., Patrinos, G.P., 2015. Economic Evaluation in Genomic Medicine. Elsevier/Academic Press, Burlington, CA, USA.

Fragoulakis, V., Mitropoulou, C., van Schaik, R.H., Maniadakis, N., Patrinos, G.P., 2016. An alternative methodological approach for cost-effectiveness analysis and decision making in genomic medicine. OMICS 20, 274–282.

Fragoulakis, V., Mitropoulou, C., Katelidou, D., van Schaik, R.H., Maniadakis, N., Patrinos, G.P., 2017. Performance ratio-based resource allocation decision making in genomic medicine. OMICS 21, 67–73.

Gullapalli, R.R., Lyons-Weiler, M., Petrosko, P., Dhir, R., Becich, M.J., LaFramboise, W.A., 2012. Clinical integration of next-generation sequencing technology. Clin. Lab. Med. 32, 585–599.

Hanahan, D., Weinberg, R.A., 2011. Hallmarks of cancer: the next generation. Cell 144, 646–674.

Kampourakis, K., Vayena, E., Mitropoulou, C., Borg, J., van Schaik, R.H., Cooper, D.N., Patrinos, G.P., 2014. Next generation pharmacogenomics: key challenges ahead. EMBO Rep. 15, 472–476.

Karageorgos, I., Giannopoulou, E., Mizzi, C., Pavlidis, C., Peters, B., Karamitri, A., Zagoriti, Z., Stenson, P., Kalofonos, H.P., Drmanac, R., Borg, J., Cooper, D.N., Katsila, T., Patrinos, G.P., 2015. Identification of cancer predisposition variants using a next generation sequencing-based family genomics approach. Hum. Genomics 9, 12.

Kilpinen, H., Barrett, J.C., 2013. How next-generation sequencing is transforming complex disease genetics. Trends Genet. 29, 23–30.

Kricka, L.J., Fortina, P., Mai, Y., Patrinos, G.P., 2011. Direct-to-consumer genetic testing: a view from Europe. Nat. Rev. Genet. 12, 670.

Lazaridis, K.N., McAllister, T.M., Babovic-Vuksanovic, D., Beck, S.A., Borad, M.J., Bryce, A.H., Chanan-Khan, A.A., Ferber, M.J., Fonseca, R., Johnson, K.J., Klee, E.W., Lindor, N.M., McCormick, J.B., McWilliams, R.R., Parker, A.S., Riegert-Johnson, D.L., Rohrer Vitek, C.R., Schahl, K.A., Schultz, C., Stewart, K., Then, G.C., Wieben, E.D., Farrugia, G., 2014. Implementing individualized medicine into the medical practice. Am. J. Med. Genet. C Semin. Med. Genet. 166, 15–23.

Logue, L.J., 2003. Genetic testing coverage and reimbursement: a provider's dilemma. Clin. Lab. Manage. Rep. 17, 346–350.

Macintyre, G., Ylstra, B., Brenton, J.D., 2016. Sequencing structural variants in cancer for precision therapeutics. Trends Genet. 32, 530–542.

Mai, Y., Mitropoulou, C., Papadopoulou, X.E., Vozikis, A., Cooper, D.N., van Schaik, R.H., Patrinos, G.P., 2014. Critical appraisal of the views of healthcare professionals with respect to pharmacogenomics and personalized medicine in Greece. Pers. Med. 11, 15–26.

Manolio, T.A., Abramowicz, M., Al-Mulla, F., Anderson, W., Balling, R., Berger, A.C., Bleyl, S., Chakravarti, A., Chantratita, W., Chisholm, R.L., Dissanayake, V.H., Dunn, M., Dzau, V.J., Han, B.G., Hubbard, T., Kolbe, A., Korf, B., Kubo, M., Lasko, P., Leego, E., Mahasirimongkol, S., Majumdar, P.P., Matthijs, G., McLeod, H.L., Metspalu, A., Meulien, P., Miyano, S., Naparstek, Y., O'Rourke, P.P., Patrinos, G.P., Rehm, H.L., Relling, M.V., Rennert, G., Rodriguez, L.L., Roden, D.M., Shuldiner, A.R., Sinha, S., Tan, P., Ulfendahl, M., Ward, R., Williams, M.S., Wong, J.E., Green, E.D., Ginsburg, G.S., 2015. Global implementation of genomic medicine: we are not alone. Sci. Transl. Med. 7, 290ps13.

Mitropoulou, C., Mai, Y., van Schaik, R.H., Vozikis, A., Patrinos, G.P., 2014. Documentation and analysis of the policy environment and key stakeholders in pharmacogenomics and genomic medicine in Greece. Public Health Genomics 17, 280–286.

Mitropoulou, C., Fragoulakis, V., Bozina, N., Vozikis, A., Supe, S., Bozina, T., Poljakovic, Z., van Schaik, R.H., Patrinos, G.P., 2015. Economic evaluation for pharmacogenomic-guided warfarin treatment for elderly Croatian patients with atrial fibrillation. Pharmacogenomics 16, 137–148.

Mitropoulou, C., Fragoulakis, V., Rakicevic, L.B., Novkovic, M.M., Vozikis, A., Matic, D.M., Antonijevic, N.M., Radojkovic, D.P., van Schaik, R.H., Patrinos, G.P., 2016. Economic analysis of pharmacogenomic-guided clopidogrel treatment in Serbian patients with myocardial infarction undergoing primary percutaneous coronary intervention. Pharmacogenomics 17, 1775–1784.

Mizzi, C., Mitropoulou, C., Mitropoulos, K., Peters, B., Agarwal, M.R., van Schaik, R.H., Drmanac, R., Borg, J., Patrinos, G.P., 2014. Personalized pharmacogenomics profiling using whole genome sequencing. Pharmacogenomics 15, 1223–1234.

Mizzi, C., Dalabira, E., Kumuthini, J., Dzimiri, N., Balogh, I., Başak, N., Böhm, R., Borg, J., Borgiani, P., Bozina, N., Bruckmueller, H., Burzynska, B., Carracedo, A., Cascorbi, I., Deltas, C., Dolzan, V., Fenech, A., Grech, G., Kasiulevicius, V., Kádaši, Ľ., Kučinskas, V., Khusnutdinova, E., Loukas, Y.L., Macek, Jr., M., Makukh, H., Mathijssen, R., Mitropoulos, K., Mitropoulou, C., Novelli, G., Papantoni, I., Pavlovic, S., Saglio, G., Setric, J., Stojiljkovic, M., Stubbs, A.P., Squassina, A., Torres, M., Turnovec, M., van Schaik, R.H., Voskarides, K., Wakil, S.M., Werk, A., Del Zompo, M., Zukic, B., Katsila, T., Lee, M.T., Motsinger-Rief, A., Mc Leod, H.L., van der Spek, P.J., Patrinos, G.P., 2016. A European spectrum of pharmacogenomic biomarkers: implications for clinical pharmacogenomics. PLoS One 11, e0162866.

Papadopoulos, P., Viennas, E., Gkantouna, V., Pavlidis, C., Bartsakoulia, M., Ioannou, Z.M., Ratbi, I., Sefiani, A., Tsaknakis, J., Poulas, K., Tzimas, G., Patrinos, G.P., 2014. Developments in

FINDbase worldwide database for clinically relevant genomic variation allele frequencies. Nucleic Acids Res. 42, D1020–D1026.

Patrinos, G.P., 2006. National and ethnic mutation databases: recording populations' genography. Hum. Mutat. 27, 879–887.

Patrinos, G.P., Brand, A., 2014. Public health genomics joins forces with the Genomic Medicine Alliance. Public Health Genomics 17, 125–126.

Patrinos, G.P., Al Aama, J., Al Aqeel, A., Al-Mulla, F., Borg, J., Devereux, A., Felice, A.E., Macrae, F., Marafie, M.J., Petersen, M.B., Qi, M., Ramesar, R.S., Zlotogora, J., Cotton, R.G., 2011. Recommendations for genetic variation data capture in emerging and developing countries to ensure a comprehensive worldwide data collection. Hum. Mutat. 32, 2–9.

Patrinos, G.P., Baker, D.J., Al-Mulla, F., Vasiliou, V., Cooper, D.N., 2013. Genetic tests obtainable through pharmacies: the good, the bad and the ugly. Hum. Genomics 7, 17.

Pavlidis, C., Lanara, Z., Balasopoulou, A., Nebel, J.C., Katsila, T., Patrinos, G.P., 2015. Meta-analysis of nutrigenomic biomarkers denotes lack of association with dietary intake and nutrient-related pathologies. OMICS 19, 512–520.

Pavlidis, C., Nebel, J.C., Katsila, T., Patrinos, G.P., 2016. Nutrigenomics 2.0: the need for ongoing and independent evaluation and synthesis of commercial nutrigenomics tests' scientific knowledge base for responsible innovation. OMICS 20, 65–68.

Payne, K., Shabaruddin, F.H., 2010. Cost-effectiveness analysis in pharmacogenomics. Pharmacogenomics 11, 643–646.

Pisanu, C., Tsermpini, E.E., Mavroidi, E., Katsila, T., Patrinos, G.P., Squassina, A., 2014. Assessment of the pharmacogenomics educational environment in southeast Europe. Public Health Genomics 17, 272–279.

Prainsack, B., Vayena, E., 2013. Beyond the clinic: "Direct-to-consumer" genomic profiling services and pharmacogenomics. Pharmacogenomics 14, 403–412.

Reydon, T.A., Kampourakis, K., Patrinos, G.P., 2012. Genetics, genomics and society: the responsibilities of scientists for science communication and education. Pers. Med. 9, 633–643.

Snyder, S.R., Mitropoulou, C., Patrinos, G.P., Williams, M.S., 2014. Economic evaluation of pharmacogenomics: a value-based approach to pragmatic decision making in the face of complexity. Public Health Genomics 17, 256–264.

Squassina, a., Severino, G., Grech, G., Fenech, a., Borg, J., Patrinos, G.P., 2012. Golden Helix pharmacogenomics days: educational activities on pharmacogenomics and personalized medicine. Pharmacogenomics 13, 525–528.

Viennas, E., Komianou, a., Mizzi, C., Stojiljkovic, M., Mitropoulou, C., Muilu, J., Vihinen, M., Grypioti, P., Papadaki, S., Pavlidis, C., Zukic, B., Katsila, T., van der Spek, P.J., Pavlovic, S., Tzimas, G., Patrinos, G.P., 2017. Expanded national database collection and data coverage in the FINDbase worldwide database for clinically relevant genomic variation allele frequencies. Nucleic Acids Res. 45, D846–D853.

Vozikis, a., Cooper, D.N., Mitropoulou, C., Kambouris, M.E., Brand, a., Dolzan, V., Fortina, P., Innocenti, F., Lee, M.T., Leyens, L., Macek, Jr., M., Al-Mulla, F., Prainsack, B., Squassina, A., Taruscio, D., van Schaik, R.H., Vayena, E., Williams, M.S., Patrinos, G.P., 2016. Test pricing and reimbursement in genomic medicine: towards a general strategy. Public Health Genomics 19, 352–363.

# Index

Printed in the United States
By Bookmasters